SECONDARY SCHOOL MATHEMATICS 1
INTERMEDIATE DIVISION

NOTES

COLES EDITORIAL BOARD

Bound to stay open

Publisher's Note

Otabind (Ota-bind). This book has been bound using the patented Otabind process. You can open this book at any page, gently run your finger down the spine, and the pages will lie flat.

ABOUT COLES NOTES

COLES NOTES have been an indispensible aid to students on five continents since 1948.

COLES NOTES are available for a wide range of individual literary works. Clear, concise explanations and insights are provided along with interesting interpretations and evaluations.

Proper use of COLES NOTES will allow the student to pay greater attention to lectures and spend less time taking notes. This will result in a broader understanding of the work being studied and will free the student for increased participation in discussions.

COLES NOTES are an invaluable aid for review and exam preparation as well as an invitation to explore different interpretive paths.

COLES NOTES are written by experts in their fields. It should be noted that any literary judgement expressed herein is just that – the judgement of one school of thought. Interpretations that diverge from, or totally disagree with any criticism may be equally valid.

COLES NOTES are designed to supplement the text and are not intended as a substitute for reading the text itself. Use of the NOTES will serve not only to clarify the work being studied, but should enhance the readers enjoyment of the topic.

ISBN 0-7740-3421-1

© COPYRIGHT 1997 AND PUBLISHED BY
COLES PUBLISHING COMPANY
TORONTO - CANADA
PRINTED IN CANADA

Manufactured by Webcom Limited
Cover finish: Webcom's Exclusive **DURACOAT**

CONTENTS

SECONDARY SCHOOL MATHEMATICS 1

(INTERMEDIATE DIVISION)

These notes have been set out in four main divisions:

1. Glossary of symbols

2. Definitions

3. Some notes and helps

4. Sample examinations and their solutions

This has been done with one purpose in mind — to provide you with a concise statement of the New Approach Mathematics as it is being handled by many teachers today.

In this period of your mathematical growth, this book is intended to give you some idea of the scope of the changing subject, mathematics, while suggesting, through lists of examinations, certain goals which you may set for yourself in one year.

GLOSSARY OF MATHEMATICAL SYMBOLS

The following consists of some of the abbreviations used in the language of mathematics. If you are to master this subject, you should have a thorough knowledge of these symbols. Learn them as you progress through the course.

Mathematical Symbol	Example
$\because$ because, since	
$\therefore$ therefore	
$+$ plus, add, positive	$4 + 5 \leftrightarrow$ Four plus (add) Five $+ 3 \leftrightarrow$ positive three
$-$ minus, subtract, negative	$6 - 2 \leftrightarrow 6$ subtract (minus) two $- 8 \leftrightarrow$ negative 8
$\times$ or $\cdot$ multiplied by	
$\div$ divided by	
$=$ is equal to	
$\neq$ is NOT equal to	
$\doteq, \stackrel{.}{=}, \simeq$ is approximately equal to	$3.01 \doteq 3$
$\equiv, \cong,$ is congruent to	$\triangle ABC \equiv \triangle DEF$

6

| $\lVert$, ~ is similar to | $\triangle\,XYZ \,\lVert\, \triangle\,PGR$ |

$>$ is greater than $5 > 1$

$\ngtr$ is NOT greater than $6 \ngtr 9$

$<$ is less than $2 < 4$

$\nless$ is NOT less than $3 \nless 1$

$\geq$ is greater than or equal to

$\ngeq$ is NOT greater than or equal to

$\leq$ is less than or equal to

$\nleq$ is NOT less than or equal to

$\lvert\ \rvert$ the absolute value of $\lvert\,7\,\rvert \longleftrightarrow$ the absolute value of 7

$\lVert$ is parallel to

$\perp$ is perpendicular to

$\angle$ the angle

$\triangle$ the triangle

$\lVert\,$gm   the parallelogram

$\{\ \}$ the set

ϵ belongs to, is a member of

$\{\ \}$ the set of $\{1,2,3,4\} \longleftrightarrow$ the set of 1,2,3 and 4

ϵ belongs to, is a member of $5 \in \{3,5,7,9,11\}$

$\{\ \}, \phi$ the null (empty) set

$N = \{1,2,3,4, \ldots\}$

$W, No = \{0,1,2,3, \ldots\}$

$I = \{\ldots, -3,-2,-1,0,1,2,3, \ldots\}$

$Q = \{\frac{a}{b} \mid a, b \in I, b \neq 0\}$

$\overline{Q}$ = the set of irrational numbers

R = the set of real numbers, $R = Q \cup \overline{Q}$

∩ = **the intersection (of two sets)**

∪ = **the union (of two sets)**

⟷ , ⟺   is equivalent to $3x = 9 \Longleftrightarrow x = 3$

⟹   implies, if . . ., then . . .

LIST OF DEFINITIONS

Definitions and specific examples have been given here with the hope that you will find greater value in this list as you see the general definition linked to numerical or diagrammatic explanations.

Word	**Definition**	**Example**
absolute value	the positive value of a number	$\lvert 3 \rvert = 3, \lvert -3 \rvert = 3$ $\lvert -\tfrac{1}{2} \rvert = \tfrac{1}{2} \lvert \tfrac{3}{4} \rvert = \tfrac{3}{4}$
acute angle	**an angle whose degree measure is between 0 and 90**	

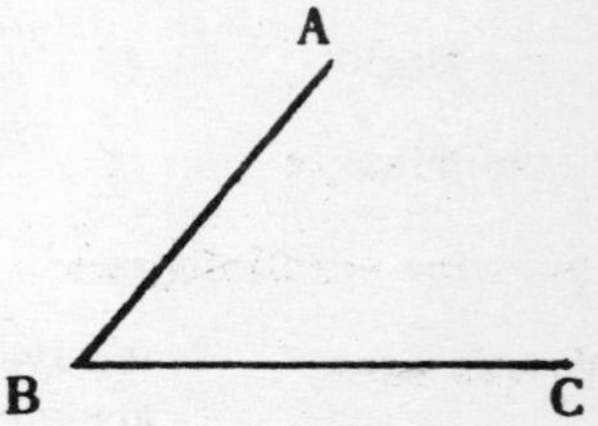

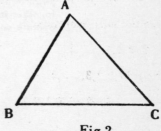

Fig. 1.

∠ABC is acute

acute angled triangle	**a triangle having each of its interior angles acute**	

Fig. 2.

△ABC is acute-angled

additive inverse   see opposites

8

adjacent angles   two angles having a
common vertex, a
common side, and
sides on opposite
sides of the common
side

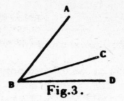

Fig.3.

∠ABC and ∠DBC are
adjacent angles

angle   two rays (arms) terminating
at a fixed point (vertex)

arc (of a circle)   a portion of the
circumference of a
circle

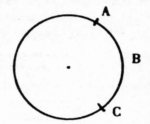

Fig.4.

ABC is an arc of the
circle

arithmetic mean   the average of a
group of numbers

the average of 5, 9 and 4

is $\dfrac{5+9+4}{3} = 6$

associative property   the grouping of
pairs of numbers
together in order to
operate on them

(1) the associative pro-
perty for addition holds
$(a + b) + c = a + (b + c)$

(2) the associative pro-
perty for multiplication
holds
$(a \times b) \times c = a \times (b \times c)$

(3) the associative pro-
perty does not hold for
subtraction or division
$(3 - 2) - 4 \neq 3 - (2 - 4)$
$(4 \div 2) \div 3 \neq 4 \div (2 \div 3)$

| base | the basic unit for counting in a system of numeration | count in the system to base 7: 1, 2, 3, 4, 5, 6, 10, 11, 12 etc. Check your text for further work in this. |

| base | that part of a power which appears as a factor when the power is "expanded" | 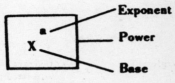 |

Fig.5.

| binomial | a two termed algebraic expression | a + 4b
7x − 5 |

| centroid (of a triangle) | the point of concurrence of the medians of a triangle | 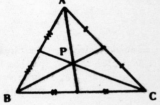 |

Fig.6.

P is the centroid of △ ABC

| chord (of a circle) | a line segment terminated at either end by points on a circle | 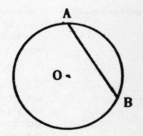 |

Fig.7.

AB is a chord of the circle with centre O

circle	the set of points equidistant from a given fixed point	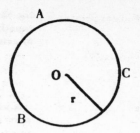

Fig.8.

a circle, each point on it being distance r from the fixed point O

circumcentre   the centre of the circumscribed circle of a triangle

circumference   the measure of the perimeter of the circle

circumscribed circle (circumcircle)   the circle which passes through the three vertices of a triangle

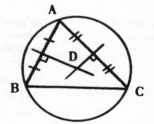

Fig.9.

D, the point of intersection of the right bisectors of the sides, is the circumcentre.

closure   that property of a set (of numbers) by which when we operate on any two of them we obtain a unique answer which is a member of the original set

$2, 3 \in N$
$2 \times 3 = 6$ (a unique result)
and $6 \in N$

collinear points	points which lie in a straight line		
commutative property	that property of an operation which permits reversal of the numbers operated on	$a + b = b + a$ $a \times b = b \times a$ $a - b \neq b - a$ $a \div b \neq b \div a$	$5 + 3 = 3 + 5$ $2 \times 3 = 3 \times 2$ $7 - 4 \neq 4 - 7$ $8 \div 2 \neq 2 \div 8$

complementary angles

a pair of angles whose sum is 90°

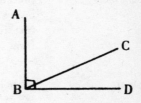

Fig. 10.

∠ABC and ∠CBD are complementary

cone (right circular)

if a stick is held with one end fixed, and the other end is moved in a circular path, we say that a cone is "generated". If the base of this cone is so cut that the cone sits "upright", it is then *right circular*

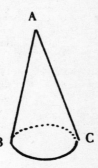

Fig. 11.

ABC is a cone.

congruent

two figures are congruent if they are alike (identical) in all respects except position

cylinder

If a stick is stood on end on a flat table and then moved along the table in a circular path (still remaining upright) we say that a circular cylinder has been "generated"

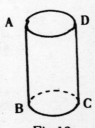

Fig. 12.

ABCD is a cylinder

deductive method	a proof, or solution based on a step-by-step development using previously stated postulates or theorems	
denominator	the number appearing below the line in a fraction in rational form	¾ is a fraction having 3 as numerator, and 4 as denominator
diagonal	a line joining any two non-adjacent vertices of a rectilineal figure (a figure bounded by straight lines)	
diameter	any chord of a circle which passes through the centre	

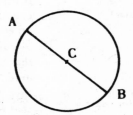

Fig. 13.

AB is a diameter — it is a chord passing through the centre C.

distributive property	that property by which we are able to multiply over addition	$4(x + y) = 4x + 4y$ $(a + b)(c + d) =$ $(a + b)c + (a + b)d$
division	the inverse of multiplication	$\frac{16}{2} = 8$ implies that $8 \times 2 = 16$
equation	a statement of equality of two numbers	$3x + 5 = 10$ is an equation

equilateral triangle	a triangle in which all three sides are equal	

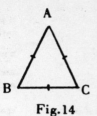

Fig. 14

Δ ABC is equilateral

equivalent equations (inequalities)	these are equations or inequalities which have the same solution set	$2x + 1 = 5$ and $2x = 4$ are equivalent equations because 2 is the root of both

exponent   that part of a power which indicates the number of times the base will appear as a factor

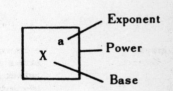

Fig. 15.

a^5 means $a \times a \times a \times a \times a$
there are 5 factors of a

exterior angle   If one arm of an angle lies along one side of a polygon, and the other arm is on an adjacent side of the polygon produced, then the angle is exterior with respect to the polygon

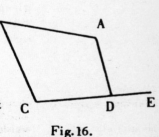

Fig. 16.

$\angle$ADE is an exterior angle

factors   numbers being multiplied or divided

	a number which will divide exactly into another is a factor of the number	6 is a factor of 42 because $\frac{42}{6}$ = 7 with no remainder

14

fraction	a fraction, in rational form, is a number $\frac{a}{b}$ where a and b are integers and $b \neq 0$	$F = \left\{ \frac{a}{b} \mid a \in N_0, b \in N \right\}$ $Q = \left\{ \frac{a}{b} \mid a, b \in I, b \neq 0 \right\}$

frustum (of a cone) that portion of a cone which remains when the "top" has been severed or the side opposite the right angle

Fig. 17.

frustum of a cone

graph a drawing which shows a relation between certain sets of numbers

graph the solution set of $3x > 2x + 5, x \in I$

i.e. $x > 5$

+1 +2 +3 +4 +5 +6 +7 +8 +9   I-line

Fig. 18.

hypotenuse the longest side of a right angled triangle or the side opposite the right angle

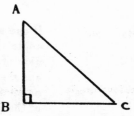

Fig. 19.

AC is the hypotenuse

identity element a number, which, when operated on by a given number, gives the given number as a result

$\because a + 0 = a$

$\therefore 0$ is the identity element for addition

$\because a \times 1 = a$

$\therefore 1$ is the identity element for multiplication

inductive method the method of drawing reasonable conclusions from repeated trials or experiments yielding non-general results

inequality	a statement that two quantities are unequal	(1) $5 > 3$ (2) $3x - 2 \leq 2x + 5$ are inequalities
inscribed circle (of a triangle)	the circle drawn tangent to the three sides of a triangle	

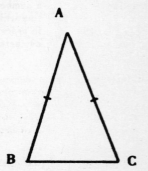

Fig. 20.

circle with centre D is
inscribed in the triangle
D is the point common to
the bisectors of the angles

integer	a "whole" number, positive or negative	(1) $I = \{ 0, \pm 1, \pm 2, \pm 3 \dots \}$, i.e. $\{ \dots -3, -2, -1, 0, +1, +3 \dots \}$
irrational number	a non-recurring, non-terminating decimal fraction	these are of two main kinds: (1) square roots such as $\sqrt{2}, \sqrt{3}, \sqrt{5}$ (2) non square roots such as π (In 1961 the value of π was worked out by machine to 100,000 places of decimals without ever repeating or recurring.)
isosceles triangle	a triangle having two equal sides	

A

B   C

Fig. 21.

$\triangle ABC$ is isosceles
because $AB = AC$

line segment	a line segment consists of two points on it, and all the points between these two points	 Fig. 22.

AB is a line segment of XY.

median	the line drawn from the mid point of one side of a triangle to the opposite vertex	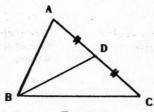 Fig. 23.

BD is a median of △ ABC

monomial	a single term algebraic expression	3x, −5a are monomials

natural numbers	a set of signless numbers used for counting	$N = \{1, 2, 3, 4, ...\}$

negative (number)	the opposite of a positive number	−3 is the opposite of +3 −3 is negative in sense

number line	a number line is a line which we draw in order to plot numbers on it	−2 −1 0 +1 +2 +3 +4 +5 I-line Fig. 24. −2 −1 0 +1 +2 +3 +4 Q-line Fig. 25. are number lines. In each case (where ambiguity may occurr) the name of the line is indicated.

numerator	the number above the line in a fraction	¾ is a fraction having 3 as <u>numerator</u> and 4 as denominator

obtuse angle   an angle whose degree measure lies between 90° and 180°

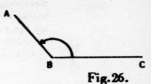

Fig. 26.

∠ ABC is an obtuse angle

obtuse angled triangle   a triangle having one of its angles obtuse

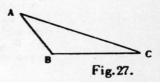

Fig. 27.

∴ ∠ ABC is obtuse, the triangle is called an obtuse angled triangle

opposite angles (vertical angles)   two angles such that each side of one is the prolongation through the vertex of a side of the other

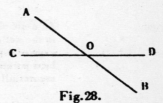

Fig. 28.

∠AOC, ∠DOB and ∠AOD, ∠COB are opposite pairs of angles

opposites   two numbers whose sum is zero

the number line is symmetrical about the zero point in that for every number to the right of it there is a corresponding negative number to the left

4 and −4 are opposites
$\frac{2}{3}$ and −$\frac{2}{3}$ are opposites

parallel lines   these are lines which do not meet, no matter how far they be produced in either direction

A ———————————— B

C ———————————— D

Fig. 29.

AB || CD is read "AB is parallel to CD".

18

| parallelogram | a quadrilateral having both pair of opposite sides parallel | 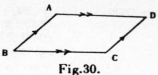 |

Fig.30.

ABCD is a || gm.

| perimeter | the total distance around a figure |

| perpendicular | a line drawn at right angles to another is said to be perpendicular to it | 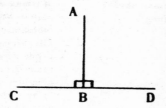 |

Fig.31.
AB is perpendicular to CD

| polygon | An n-sided polygon is the set of n line segments determined by n points which are not all collinear. |

| polynominal | a sum of terms | $3x, 2x + 5y, 2y + 3a + b,$ $6x^3 + 7x^2 - 8x - 13$ are <u>all</u> polynominals. Special names are given to the first three:
monomial — one term
binomial — two terms
trinomial — three terms |

| positive | that property of every number lying to the right of zero on a number line |

$$-3 \ -2 \ -1 \ \ 0 \ \ +1 \ +2 \ +3 \ +4 \quad \text{R-line}$$

Fig.32.

| power | the value assigned to a number with an exponent | 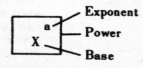 |

Fig.33.

prime number

an integer which has no integral factor except unity and itself. 1 is usually excluded.

A few small primes are 2, 3, 5, 7, 11.

prism

a many-sided solid figure, each of whose sides, called <u>lateral</u> surfaces, are parallelograms. (They may be rectangles, but we consider a rectangle to be a special case of parallelogram.)

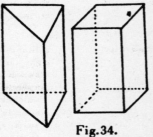

Fig. 34.

Examples of upright or <u>right</u> prisms the base of a prism may be any rectilineal figure.

pyramid

a solid figure having a base which is a rectilineal figure, and whose other sides are triangles with a common vertex

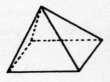

Fig. 35.

the example shows a rectangular pyramid

quadrilateral

a polygon having four sides

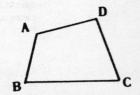

Fig. 36

ABCD is a quadrilateral

radius   the length of the line joining the centre of a circle to any point on the circle

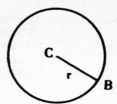

Fig.37.
CB is a radius of the circle

ratio   an ordered set of numbers, each of which is the measure of a quantity, and is measured in the same unit

3: 5: 7: 8 is a ratio read "3 is to 5 is to 7 is to 8".
a: b is a two termed ratio and we operate on it as we would on the fraction $\frac{a}{b}$ as long as $b \neq o$

Note, however that 3:0 is a perfectly valid ratio, but may not be considered as a meaningful fraction.

rational number   a number of the form $\frac{a}{b}$ where $b \neq o$

$Q = \{ \frac{a}{b} \,|\, a, b \in I, b \neq o \}$
Decimal fractions which terminate or recur may be reduced to the form $\frac{a}{b}$

ray   a line having one fixed point, the remainder of the points lying to the same side of the fixed point

A   B

Fig.38.
AB is a ray, the point A being fixed.

reciprocal   two numbers are reciprocals if their product is one

2 and ½ are reciprocals because $2 \times ½ = 1$
-5 and $-\frac{1}{5}$ are reciprocals because $-5x -\frac{1}{5} = 1$

rectilineal   straight lined

rectangle	a parallelogram having one right angle	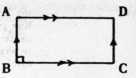

Fig. 39.

ABCD is a rectangle. Note that if it has <u>one</u> right angle then it <u>must</u> have four. <u>But stick to the definition.</u>

regular polygon	a polygon having all sides equal and all angles equal	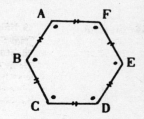

Fig. 40.

ABCD is a regular hexagon

replacement set	the set of numbers which may replace a variable.	$x + 2 > 4,\ x \in \{1,2,3,4,5,6,7\}$ $\{1,2,3,4,5,6,7\}$ is the replacement set

rhombus	a parallelogram having four equal sides	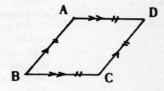

Fig. 41.

ABCD is a rhombus.

right angle	an angle whose degree measure is 90°	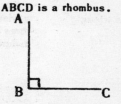

Fig. 42.

$\angle ABC = 90°$

right angled triangle — a triangle having one of its angles a right angle

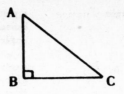

Fig.43.

$\triangle$ABC is a right angled triangle

right bisector — a line which is at right angled to another, and meets it at its mid point

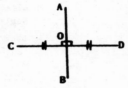

Fig.44.

AB is the right bisector of CD

root — a number is the root of an equation if when it replaces the variable in the left hand number and the right hand number, the LHN equals the RHN.

$2x = x + 3, x \in N$

$x = 3$

LHN $= 2x$ RHN $= x + 3$

     $= 2 \times 3$ $= 3 + 3$

     $= 6$ $= 6$

$\therefore$ 3 is the root of the equation

scalene triangle — a triangle having three sides of differing length

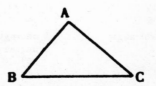

Fig.45.

AB $\neq$ BC $\neq$ CA, therefore $\triangle$ABC is scalene

		Number	Scientific Notation
scientific notation | a method of writing a number adopted by scientists as easy to understand. A decimal point is placed after the first meaningful figure, and a suitable power of 10 is inserted as a factor to locate the decimal point correctly. | 348 $= 3.48 \times 10^2$
 25.37 $= 2.537 \times 10$
 53000 $= 5.3 \times 10^4$
 53000 $= 5.300 \times 10^4$
 .00425 $= 4.25 \times 10^{-3}$
 53000 may be accurate to two, three, four or five figures. Only the accurate figures are written in scientific notation. Your teacher might tell you the meaning of 10^{-3}. |

sector   a region of a circle bounded by two radii of the circle and one of the arcs which they cut off

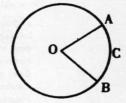

Fig. 46.

AOBC is a sector
ACB is called the <u>minor arc.</u>

sector angle   an angle at the centre of a circle standing on an arc of the circle

In fig. 46, $\angle AOB$ is a sector angle.

segment (of a circle)   the area between a chord and an arc subtended by the chord

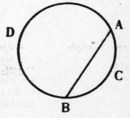

Fig. 47.

ABC is the minor segment
ABD is the major segment

semi circle   the area bounded by the longest chord of a circle and its subtending arc. The longest chord of a circle is its diameter.

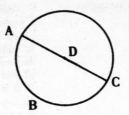

Fig. 48.

ABC is a semi circle if D is the centre of the circle.

set   a group of elements having some common stated characteristic. We deal with sets of numbers in mathematics

You should be familiar with the following sets of numbers:

$N = \{1, 2, 3, 4 ...\}$

$N_0 = \{0, 1, 2, 3 ...\}$

$F = \{\frac{a}{b} \mid a, b \in N\}$

$F_0 = \{\frac{a}{b} \mid a \in N_0, b \in N\}$

$I = \{0, \pm 1, \pm 2, \pm 3 ...\}$

$Q = \{\frac{a}{b} \mid a, b \in I, b \neq 0\}$

R = the set of real numbers

set-builder   the short form of showing a relation, or a required set of numbers

$T = \{x \mid 3x - 2 > x + 6, x \in R\}$ is read "T is the set of x such that $3x - 2 > x + 6$ and x is a member of R" The glossary of symbols will give you the meanings of specific symbols used above.

similar triangles	triangles whose angles are equal and whose corresponding sides form equal ratios, are said to be similar	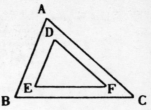 **Fig. 49.**

$\Delta ABC \mid\mid\mid \Delta DEF$

it would appear that of two similar figures, one is a "model" of the other

solution set	the set of the number(s) which is the solution of the equation or inequation	$5x = 10$ $x = 2$ $\therefore$ the solution set is $\{2\}$
sphere	The set of points, in space, at a given distance from a fixed point.	a baseball is a sphere
square	a rectangle having all sides equal	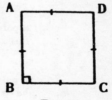 **Fig. 50.** **ABCD is a square**
square root	one of the two identical factors whose product is the number	$3 \times 3 = 9$ $\therefore \sqrt{9} = 3$
subset	the set of some or all of the members of a given set	(1) $\{3,8\}$ is a subset of $\{2,3,5,8\}$ (2) ϕ is a subset of $\{2, 3, 5, 8\}$ because we could choose **NO members** (3) $\{2, 3, 5, 8\}$ is a subset of $\{2, 3, 5, 8\}$; every set is a subset of itself

supplementary angles	two angles whose sum has a degree measure of 180 are supplementary	

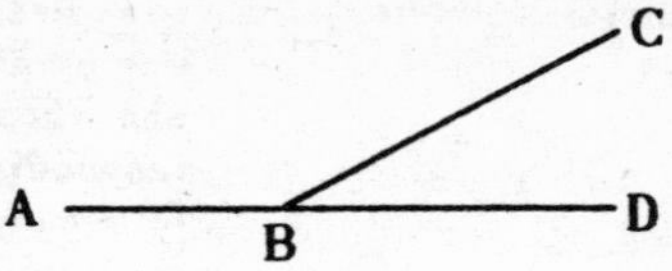

Fig.51.

$\angle ABC + \angle CBD = 180°$

terms	numbers being added or subtracted	$5x + 2y - 3z$ $5x, 2y$ and $3z$ are terms

trapezoid	a quadrilateral having one pair of opposite sides paralled	

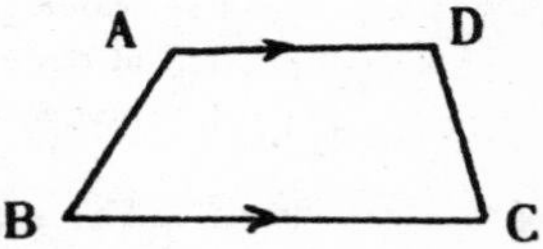

Fig.52.

ABCD is a trapezoid

triangle	a polygon having three sides

trinomial	a polynomial having three terms	$3a + 2b - 4c, x^2 - 2xy + y^2$ are trinomials

unity	a word used to indicate the numeral 1

vertex	the common point of two rays of an angle

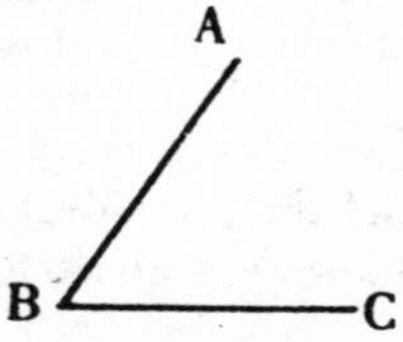

Fig.53.

B is the vertex of the $\angle ABC$

SOME NOTES AND AIDS

This section is intended to serve as a brief summary of topics which are important to the students progress in mathematics. In most cases worked examples are shown and then followed by unworked examples accompanied by answers. It is strongly suggested that the student attempt all of these examples in order to make best use of the notes and his own time.

ALGEBRA

ORDER OF OPERATIONS

There are four fundamental operations in mathematics; addition, subtraction, multiplication and division. Since many problems may involve any or all of these operations, it is important to decide on the order in which the operations are to be carried out. Therefore

1. Operations within brackets (if any) are to be done first.
2. Multiplication and division are then done in the order that they appear reading from left to right.
3. Addition and subtraction are then finally done.

e.g.
$$\text{i)} \quad (5+4) + 6 \times 3 = 9 + 6 \times 3$$
$$= 9 + 18$$
$$= 27$$

$$\text{ii)} \quad (4 \times 3 - 8) - 9 \div 3 = (12 - 8) - 9 \div 3$$
$$= 4 - 9 \div 3$$
$$= 4 - 3$$
$$= 1$$

$$\text{iii)} \quad 15 - 4 \times 2 = 15 - 8$$
$$= 7$$

$$\text{iv)} \quad 12 \times 3 \div 4 = 36 \div 4$$
$$= 9$$

Note: When doing a question of this type, *scan the whole calculation* and then decide which operation must be done first. Continue doing this at each stage of the calculation.

Examples

1. $8 \times 4 \div 2$ **Answer** 16
2. $9 + (5 - 2)$ 12
3. $(6 \times 2 - 1) - 2 \times 4$ 3

4. $12 - 16 \div 2$ **4**

5. $2 \times (7 - 2 \times 3) - 1$ **1**

FUNDEMENTAL OPERATIONS WITH INTEGERS

$$I = \{ \ldots, -3, -2, -1, 0, 1, 2, 3, \ldots \}$$

1. Addition   (term) + (term) = sum

  i) The sum of two positive numbers is a positive number

  e.g. $(+3) + (+8) = +11$

  ii) The sum of two negative numbers is a negative number

  e.g. $(-2) + (-7) = -9$

  iii) To add two numbers of unlike (opposite) sign, 1. determine the absolute value of each of the numbers, 2. subtract the smaller absolute value from the larger absolute value and then 3. prefix this result with sign of the number which had the larger absolute value

  e.g. $-6 + 2 = -(6 - 2) = -4$

  $3 + (-5) = -(5 - 3) = -2$

  $-7 + 10 = +(10 - 7) = +3$

2. Subtraction   (minuend) − (subtrahend) = difference

To subtract integers, add the opposite of the subtrahend.

e.g. $2 - (-5) = 2 + (+5) = +7$

$-6 - 3 = -6 + (-3) = -9$

$4 - 7 = 4 + (-7) = -3$

3. Multiplication and Division   (factor) × (factor) = product

                                       (dividend) ÷ (divisor) = quotient

  i) The product or quotient of two numbers of like signs is positive

  e.g. $(+4) \times (6) = +24$ $(-8) \times (-2) = +16$

  $(+9) \div (+3) = +3$ $(-10) \div (-5) = +2$

  ii) The product or quotient of two numbers of unlike signs is negative.

  e.g. $(-4) \times (6) = -24$ $(-7) \times (3) = -21$

  $(-9) \div (+3) = -3$ $(12) \div (-4) = -3$

Note: (a) When adding a series of integers, add all of the positive numbers, then add all of the negative numbers and finally add these two results for the sum.

e.g. $(-8) + 4 + (-6) + (-11) + 5$

$= (-8) + (-6) + (-11) + (4) + (5)$

$= -25 + 9$

$= -16$

(b) When multiplying a group of positive and negative integers, count the number of negative factors; if this result is even, then the sign of the product is positive and if the result of the count is odd, then the sign of the product is negative

e.g. $(-2) (3) (-1) (4) = +24$ The count of the negative factors $[(-2)$ and $(-1)]$ is even (two)

$(-5) (-1) (3) (-2) = -30$ The count of the negative factors $[(-5), (-1)$ and $(-2)]$ is odd (three)

Examples

1. $5 + (-6) + 3 - 7$ **Answer** -5 See your text
2. $9 - 11 + 3 + 2 - 4$ -1 book for more
3. $-16 \div 4 + (-3) (-2)$ 2 examples and
4. $(-1) (-2) (5) (-3) (4)$ -120 make sure that you are confident and proficient in this area

OPERATIONS WITH ZERO

Addition and Subtraction

1. When zero is added to any given number, the sum is that number

    e.g. $4 + 0 = 4$, $0 + (-11) = -11$, $a + 0 = a = 0 + a$

2. When zero is subtracted from any given number, the difference is that number.

    e.g. $7 - 0 = 7$, $-3 - 0 = -3$, $a - 0 = a$

Note: $2 - 0 \neq 0 - 2$

$2 - 0 = 2$ but $0 - 2 = 0 + (-2) = -2$

Multiplication

When any number is multiplied by zero, the product is zero

e.g. $6 \times 0 = 0$, $0 \times (-8) = 0$, $12 \times 7 \times 4 \times 0 = 0$, $a \times 0 = 0 = 0 \times a$

Division

i) When zero is divided by any number except zero, the quotient is zero.

e.g. $0 \div 2 = 0$, $0 \div (-11) = 0$, $0 \div a = 0, a \neq 0$

ii) When any number (except zero) is divided by zero, the quotient is undefined.

e.g. $5 \div 0$ is undefined , $(-10) \div 0$ is undefined ,

$a \div 0$ is undefined

Note: Stating that the quotient is undefined means that there is no number which is a satisfactory (correct) answer.

iii) When zero is divided by zero, the quotient is meaningless.

e.g. $0 \div 0$ is meaningless

Note: Stating that the quotient is meaningless means that any number is a satisfactory answer and this non-uniqueness is not desirable.

Examples

1.	$5 \times 0 + 6$	**Answer** 6
2.	$0 \div 4 + 0 \div (-3)$	0
3.	$1146 \times 58 \times 7 \times 0$	0
4.	$124 \div 0$	Undefined

OPERATIONS WITH FRACTIONS $\quad F = \{\frac{a}{b} \mid a, b \in No, b \neq o\}$

1. Addition and Subtraction

To add or subtract fractions, they must have *common (like) denominators*. Thus if we wish to add or subtract fractions, we must express each fraction in equivalent form with common denominators. To accomplish this we employ the *principle of equivalent* fractions which states that *multiplying or dividing the numerator and demoninator by the same number* produces a fraction different in appearance but having the same number value as the original fraction.

e.g. $\frac{1}{2} = \frac{1 \times 3}{2 \times 3} = \frac{3}{6}$, $\frac{4}{5} = \frac{4 \times 7}{5 \times 7} = \frac{28}{35}$, $\frac{8}{12} = \frac{8 \div 4}{12 \div 4} = \frac{2}{3}$

Therefore

$$\frac{1}{2} + \frac{2}{3} = \frac{3}{6} + \frac{4}{6} \qquad\qquad \frac{7}{12} - \frac{1}{3} = \frac{7}{12} - \frac{4}{12}$$

Then the numerators are added (or subtracted)

$$= \frac{3 + 4}{6} \qquad\qquad = \frac{7 - 4}{12}$$

$$= \frac{7}{6} \qquad\qquad = \frac{3}{12}$$

Then reduce your result to lowest terms (if neccessary)

$$= \frac{3 \div 3}{12 \div 3} = \frac{1}{4}$$

Note: i) It is advisable in establishing a common denominator, to determine the lowest common denominator. (L.C.D.) The lowest common denominator is the smallest number that *each* of the denominators will divide into evenly (i.e. leave no remainder)

ii) It is helpful to change any mixed numbers $(2\frac{2}{3})$ to improper fractions $(\frac{8}{3})$ before proceeding.

e.g. 1. $1\frac{1}{3} - \frac{1}{2} = \frac{4}{3} - \frac{1}{2} = \frac{8}{6} - \frac{3}{6} = \frac{5}{6}$

2. $2\frac{1}{2} + 1\frac{1}{5} = \frac{5}{2} + \frac{6}{5} = \frac{25}{10} + \frac{12}{10} = \frac{37}{10} = 3\frac{7}{10}$

3. $\frac{2}{3} - \frac{1}{4} + \frac{3}{8} = \frac{16}{24} - \frac{6}{24} + \frac{9}{24} = \frac{19}{24}$

See your text book for further examples

Multiplication

To multiply fractions, express any mixed numbers in improper fraction form, then multiply numerators and multiply denominators.

e.g. $1\frac{1}{2} \times \frac{2}{5} = \frac{3}{2} \times \frac{2}{5} = \frac{6}{10} = \frac{3}{5}$

$\frac{7}{8} \times \frac{4}{5} = \frac{7 \times \overset{1}{4}}{\underset{2}{8} \times 5} = \frac{7}{10}$

this simplification step is sometimes called cancellation, but in fact the numerator and denominator have been divided by the same number (4) which changes the appearance but not the value of the fraction.

$\frac{3}{8} \times 4 = \frac{3}{8} \times \frac{4}{1} = \frac{3 \times \overset{1}{4}}{\underset{2}{8} \times 1} = \frac{3}{2} = 1\frac{1}{2}$

Division

To divide fractions, express any mixed numbers in improper fraction form, take the reciprocal of the divisor (invert the divisor) and multiply.

e.g. $\dfrac{4}{5} \div \dfrac{2}{3} = \dfrac{\overset{2}{\cancel{4}}}{5} \times \dfrac{3}{\underset{1}{\cancel{2}}} = \dfrac{6}{5}$

$4\dfrac{1}{2} \div 1\dfrac{1}{4} = \dfrac{9}{2} \div \dfrac{5}{4} = \dfrac{9}{\underset{1}{\cancel{2}}} \times \dfrac{\overset{2}{\cancel{4}}}{5} = \dfrac{18}{5} = 3\dfrac{3}{5}$

$1\dfrac{1}{6} \div 2 = \dfrac{7}{6} \div \dfrac{2}{1} = \dfrac{7}{6} \times \dfrac{1}{2} = \dfrac{7}{12}$

Examples

1. $\dfrac{3}{7} \times \dfrac{14}{9}$ **Answer** $\dfrac{2}{3}$

2. $2\dfrac{2}{3} \times \dfrac{9}{4}$ 6

3. $\dfrac{9}{10} \div \dfrac{2}{5}$ $\dfrac{9}{4}$

4. $1\dfrac{3}{8} \div 3$ $\dfrac{11}{24}$

OPERATIONS WITH RATIONAL NUMBERS

$$Q = \{\tfrac{a}{b} \mid a, b \in I, b \neq 0\}$$

The set of rational numbers can be thought of as a welding together of the set of integers and the set of fractions. Thus to operate with rational numbers, we will unite the rules and methods already learned for integers and fractions.

Equivalent Rational Numbers

$\dfrac{4}{-5} = \dfrac{4 \times (-1)}{-5 \times (-1)} = \dfrac{-4}{5}$, $\dfrac{-3}{-8} = \dfrac{-3 \times (-1)}{-8 \times (-1)} = \dfrac{3}{8}$

$-\dfrac{2}{3} = \dfrac{-2}{3}$, $-1\dfrac{1}{2} = \dfrac{3}{2} = \dfrac{-3}{2}$, $-7 = \dfrac{-7}{1}$

Note: When the sign is "in the middle" (i.e. not associated with the numerator or denominator), then the sign may be put with either the numerator or the denominator *but not both*. As demonstrated in the above examples it is generally more useful to associate the sign with the numerator.

$-\dfrac{7}{8} = \dfrac{-7}{8}$ is more useful than $-\dfrac{7}{8} = \dfrac{7}{-8}$

Addition and Subtraction

1. $\dfrac{4}{5} + \dfrac{1}{-2} = \dfrac{-4}{5} + \dfrac{-1}{2} = \dfrac{-8}{10} + \dfrac{-5}{10} = \dfrac{-13}{10} = -1\dfrac{3}{10}$

2. $\dfrac{-3}{-8} + \dfrac{-1}{4} = \dfrac{3}{8} + \dfrac{-1}{4} = \dfrac{3}{8} + \dfrac{-2}{8} = \dfrac{1}{8}$

3. $\dfrac{5}{6} - \dfrac{1}{-3} = \dfrac{5}{6} - \dfrac{-1}{3} = \dfrac{5}{6} - \dfrac{-2}{\cdot 6} = \dfrac{5-(-2)}{6} = \dfrac{5+2}{6} = \dfrac{7}{6}$

4. $\dfrac{1}{-2} - \dfrac{-3}{-4} = \dfrac{-1}{2} - \dfrac{3}{4} = \dfrac{-2}{4} - \dfrac{3}{4} = \dfrac{-2-3}{4} = \dfrac{-2+(-3)}{4} = -\dfrac{5}{4}$

Note: In each of the above examples, notice that the first step was to express rational numbers involved in equivalent form but with *positive denominators*..

Multiplication and Division

1. $-\dfrac{2}{3} \times \dfrac{7}{-8} = \dfrac{-\overset{-1}{2}}{3} \times \dfrac{7}{\underset{-4}{-8}} = \dfrac{-7}{-12} = \dfrac{7}{12}$

2. $-1\dfrac{4}{5} \times \dfrac{10}{3} = -\dfrac{9}{5} \times \dfrac{10}{3} = \dfrac{-\overset{3}{\cancel{9}}}{\cancel{5}_1} \times \dfrac{\overset{2}{\cancel{10}}}{\cancel{3}_1} = -6$

3. $1\dfrac{1}{3} \div -\dfrac{5}{6} = \dfrac{4}{3} \div \dfrac{-5}{6} = \dfrac{4}{\cancel{3}_1} \times \dfrac{\overset{2}{\cancel{6}}}{-5} = \dfrac{8}{-5} = -\dfrac{8}{5}$

4. $-\dfrac{7}{8} \div -3\dfrac{1}{4} = \dfrac{-7}{8} \div \dfrac{-13}{4} = \dfrac{-7}{\cancel{8}_2} \times \dfrac{\overset{1}{\cancel{4}}}{-13} = \dfrac{-7}{-26} = \dfrac{7}{26}$

Note: In the "cancellation" step, be very careful of signs

Examples

1. $-\dfrac{1}{4} + \dfrac{2}{3} - \dfrac{1}{-6}$ **Answer** $\dfrac{7}{12}$

2. $1\dfrac{1}{5} - \dfrac{7}{10} - \dfrac{1}{2}$ 0

3. $4 - 5\dfrac{1}{3} + \dfrac{-1}{6}$ $\qquad\qquad -\dfrac{3}{2}$

4. $\left(-\dfrac{7}{3}\right)\left(\dfrac{9}{4}\right)\left(-\dfrac{8}{5}\right)$ $\qquad\qquad 8\dfrac{2}{5}$

5. $\dfrac{1}{3} \div -1\dfrac{4}{5}$ $\qquad\qquad -\dfrac{5}{27}$

6. $\left(-\dfrac{4}{9}\right) \div \left(\dfrac{2}{3}\right)(-2)$ $\qquad\qquad \dfrac{4}{3}$

SIMPLIFICATION OF ALGEBRAIC EXPRESSIONS

An algebraic expression is a mathematical phrase involving numbers, signs of operation and variables.

e.g. $3a - 2b + 8$, a and b are variables

A variable is a letter representing any number within a given set of numbers. This given set of numbers is called the replacement set for the variable.

e.g. $2n - 1$, $n \in N$ n is the variable and N is the replacement set.

If $n = 1$, then $2n - 1 = 1$
If $n = 2$, then $2n - 1 = 3$ $\left.\begin{array}{l}\\\\\\\end{array}\right\}$ $\therefore$ $2n - 1$ represents any and all
If $n = 3$, then $2n - 1 = 5$ odd numbers.
etc.

Your ability to work effectively and confidently with algebraic expressions will be invaluable in your future mathematics career.

ADDITION AND SUBTRACTION IN ALGEBRAIC EXPRESSIONS

In adding or subtracting algebraic quantities, *only like terms may be added or subtracted*. Like terms are terms which have identical letter parts.

e.g. 5a and −6a are like terms

8bc and 12bc are like terms

but 2xy and 3z are unlike terms

i) $5a - 2b + 3a - 4b = 5a + 3a - 2b - 4b = 8a - 6b$

ii) $6m^2 + 2n + m^2 - n = 6m^2 + m^2 + 2n - n = 7m^2 + n$

iii) $2xy + x^2 - 5xy + x = 2xy - 5xy + x^2 + x = -3xy + x^2 + x$

Note: In example iii) x^2 and x are not like terms since x^2 means $x \cdot x$ and thus the two terms do not have identical letter parts.

Examples

1. $3x - 2y + x + 5y$ **Answers** $4x + 3y$
2. $4ab - 2c + 5ab - d$ $9ab - 2c - d$
3. $7y^2 + 8xy - y^2 - 10xy + y$ $6y^2 - 2xy + y$

MULTIPLICATION AND DIVISION OF ALGEBRAIC EXPRESSIONS

Multiplication

$x^4 = x \cdot x \cdot x \cdot x$ 4 is the index and x^4 is called the fourth power of x.
x is the base

$-5ab = -5 \times a \times b$, $4m^2n = (4)\,(m)\,(m)\,(n)$
$= (-5)\,(a)\,(b)$

  i) $(4a)\,(-2c) = (4)\,(-2)\,(a)\,(c) = -8ac$

 ii) $(-10xy)\,(-3z) = (-10)\,(-3)\,(x)\,(y)\,(z) = 30xyz$

iii) $x^3 \cdot x^2 = \underbrace{(x)\,(x)\,(x) \cdot (x)\,(x)}_{\text{five factors of x}} = x^5$ (i.e. x^{3+2})

Example iii) demonstrates that when *multiplying* algebraic quantities with *like bases*, you *retain the common base* and *add the indices*.

 iv) $(4a^4)\,(-3a) = (4)\,(-3)\,(a^4)\,(a^1) = -12a^5$

  v) $(x^2y)\,(-2xy^3) = (-2)\,(x^2)\,(x)\,(y)\,(y^3) = -2x^3y^4$

Division

  i) $x^5 \div x^2 = \dfrac{x^5}{x^2} = \dfrac{x \cdot x \cdot x \cdot \cancel{x} \cdot \cancel{x}}{\cancel{x} \cdot \cancel{x}} = x^3$ (i.e. x^{5-2})

The above example i) demonstrates that when *dividing* algebraic quantities with *like bases*, you *retain the common base* and *subtract the indices*.

 ii) $12a^6 \div 4a^2 = 3a^{6-2} = 3a^4$

iii) $\dfrac{-18x^3y}{9xy} = -2x^{3-1} = -2x^2$ Note: The two factors of y divide out i.e. $\dfrac{y}{y} = 1$

 iv) $20m^4n^2p^3 \div 5mn^2p = 4m^3p^2$ Note: $\dfrac{n^2}{n^2} = 1$

Examples

1. $(5ab^2c)(-3abc^3)$ **Answer** $-15a^2b^3c^4$

2. $(3x^3)(-2y)(4xy^2)$ $-24x^4y^3$

3. $(12m^3n) \div (6mn)$ $2m^2$

4. $(-9x^3yz) \div (-3x)$ $3x^2yz$

5. $(2m^3np)(-8mn^2) \div (4mnp)$ $-4m^3n^2$

6. $(-4ab)^2$ $16a^2b^2$

OPERATIONS WITH POLYNOMIALS

A polynomial is an algebraic expression with any number of. terms.

A polynomial of one term is called a monomial e.g. $-4b$

A polynomial of two terms is called a binomial e.g. $x + 3y$

A polynomial of three terms is called a trinomial e.g. $4a - 2b + 5c$

ADDITION AND SUBTRACTION OF POLYNOMIALS

Addition

i) $4x + 2y - 3z$
  $+ \ x \ - \ y + 2z$
  $\overline{5x + \ y - \ z}$

ii) $(5a + 2b - 7c) + (3a - 6b + 2c)$
    $= 8a - 4b - 5c$

Subtraction

i) $7m + 2n - \ p$
  $\ \ m + 3n - 4p$
  $\overline{6m - \ n + 3p}$

ii) $(8x - y + 3z) - (2x + y + 5z)$
    $= 6x - 2y - 2z$

Note: Be very careful in subtracting polynomials that you change the sign of *each term* in the subtrahend and then add. Check your difference by adding it to the subtrahend and comparing the result with the minuend.

MULTIPLICATION OF POLYNOMIALS (EXPANSION)

In multiplying polynomials we employ the distributive property of numbers

$a(b + c) = a \cdot b + a \cdot c$ the factor a multiplies
                                           each term in the bracket

$2 (3 + 5) = 2 \cdot 3 + 2 \cdot 5 = 6 + 10 = 16$

$2x (3y + 5z) = 2x \cdot 3y + 2x \cdot 5z$
$$= 6xy + 10xz$$

expanding

$a (b + c) = a \cdot b + a \cdot c$

factoring

i) $3a (2a + b - 5c) = (3a) (2a) + (3a) (b) + (3a) (-5c)$
$$= 6a^2 + 3ab - 15ac$$

ii) $-3x (4x - 2y) = (-3x) (4x) + (-3x) (-2y)$
$$= -12x^2 + 6xy$$

iii) $(x + 2) (x + 3) = (x + 2) (x) + (x + 2) (3)$
$$= x^2 + 2x + 3x + 6$$
$$= x^2 + 5x + 6$$

The above example iii) is an example of the product of two binomials. Since such products occur frequently, a more streamlined approach is advisable.

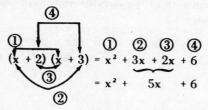

$$(x + 2) (x + 3) = x^2 + 3x + 2x + 6$$
$$= x^2 + 5x + 6$$

①   product of the *first* terms in each bracket   **F**

②   product of the *outside* terms   **O**

③   product of the *inside* terms   **I**

④   product of the *last* terms in each bracket   **L**

iv) $(a + 5) (a - 2) = a^2 - 2a + 5a - 10$
$$= a^2 + 3a - 10$$

v) $(2m - 1) (m - 3) = 2m^2 - 6m - m + 3$
$$= 2m^2 - 7m + 3$$

vi) $(b - 4)^2 = (b - 4) (b - 4) = b^2 - 4b - 4b + 16$
$$= b^2 - 8b + 16$$

vii) $(x + 2y) (x - 3y) = x^2 - 3xy + 2xy - 6y^2$
$$= x^2 - xy - 6y^2$$

viii) $(a - 2)(2a - b + 3) = (a - 2)(2a) + (a - 2)(-b) + (a - 2)(3)$
$$= 2a^2 - 4a - ab + 2b + 3a - 6$$
$$= 2a^2 - a - ab + 2b - 6$$

Examples

1. $-5a(2b + 3c - 1)$ **Answer** $-10ab - 15ac + 5a$

2. $x(2x - y) - x(x + y)$ $x^2 - 2xy$

3. $(2a - b)(3a + b)$ $6a^2 - ab - b^2$

4. $(x + 5)^2$ $x^2 + 10x + 25$

5. $(3m - 2)(m - 1)$ $3m^2 - 5m + 2$

6. $(2y - 7)^2$ $4y^2 - 28y + 49$

7. $(3b - 1)(2a - b + 1)$ $6ab - 3b^2 + 4b - 2a - 1$

SPECIAL PRODUCT

$(x + 3)(x - 3) = x^2 - 3x + 3x - 9$
$$= x^2 - 9$$

In this example, the products of the outside and inside terms ($-3x$ and $3x$) add to zero.

i) $(y + 2)(y - 2) = y^2 - 2y + 2y - 4$ ii) $(2a - 4b)(2a + 4b)$
$$= y^2 - 4$$
$$= 4a^2 + 8ab - 8ab - 16b^2$$
$$= 4a^2 - 16b^2$$

Examples

1. $(b - 1)(b + 1)$ **Answer** $b^2 - 1$

2. $(3a - 5)(3a + 5)$ $9a^2 - 25$

3. $(x + 10)(x - 10)$ $x^2 - 100$

4. $(y + 4x)(y - 4x)$ $y^2 - 16x^2$

FACTORING POLYNOMIALS

$3a(2b + 5) = 6ab + 15a$ $\therefore$ $6ab + 15a = 3a(2b + 5)$

$\underrightarrow{\text{expanding}}$ $\underrightarrow{\text{factoring}}$

In view of the previous section (expanding polynomials), factoring can be thought of as a problem where you are given the answer and asked to find the question which produced that answer.

COMMON FACTOR

A common factor of a group of terms is that number or algebraic quantity which will *divide evenly into each of the terms*.

$4a - 6b + 8c$, the common factor is 2

∴ $4a - 6b + 8c = 2(2a - 3b + 4c)$, the common factor 2 was "removed" from each term.

Note: Always check your result by expanding your answer and comparing it with the original polynomial.

It is advisable in searching for the common factor, to look for the largest possible common factor.

i) $10a^2 - 5ab = 5a(2a - b)$

ii) $12x^2 - 6xy \div 3x = 3x(4x - 2y + 1)$

iii) $4ab^2 + 8ab - 16a^2b = 4ab(b + 2 - 4a)$

Examples (Check each result by expanding)

1. $6a - 18b$ **Answer** $6(a - 3b)$

2. $12x^2 - 6x + 4xy$ $2x(6x - 3 + 2y)$

3. $a^3 + a^2 - a$ $a(a^2 + a - 1)$

4. $9m^2 - 21mn + 6m$ $3m(3m - 7n + 2)$

FACTORING SPECIAL TRINOMIALS

Recall $(x + 2)(x + 3) = x^2 + 5x + 6$

∴ $x^2 + 5x + 6$ can be written in factored form as

$(x + 2)(x + 3)$

$$a^2 + 7a + 12 = (\quad)(\quad)$$

Steps 1. Determine from the *first term* in the trinomial (a^2), what the first term in each bracket will be.

$a^2 + 7a + 12 = (a \quad)(a \quad)$

2. *Observe the last term* in the trinomial $(+12)$ and list the pairs of numbers which when multiplied will give this term.

4	−4	6	−6	12	−12
3	−3	2	−2	1	−1

3. *Observe the numerical coefficient of the middle term* of the trinomial (i.e. the number multiplying the letter) and choose from the above pairs of numbers that pair which will *add* to this number

$4 + 3 = 7$ $\therefore$ 4 and 3 are correct

$\therefore a^2 + 7a + 12 = (a + 4)(a + 3)$

4. *Check your result* by expanding and comparing with the original trinomial.

i) $x^2 - 2x - 15 = (x - 5)(x + 3)$

	√		
5	−5	15	−15
−3	3	−1	1
2	−2	14	−14

ii) $y^2 - 5y + 6 = (y - 3)(y - 2)$

		√	
3	6	−3	−6
2	1	−2	−1
5	7	−5	−7

Examples   (Check your result by expansion)

1. $a^2 + 3a + 2$ **Answer** $(a + 2)(a + 1)$

2. $x^2 - 4x - 5$ $(x - 5)(x + 1)$

3. $y^2 - 13y + 30$ $(y - 3)(y - 10)$

4. $6^2 - 2b - 24$ $(b - 6)(b + 4)$

5. $m^2 - 9m + 14$ $(m - 7)(m - 2)$

SPECIAL CASE (DIFFERENCE OF SQUARES)

Recall $(x - 2)(x + 2) = x^2 - 4$, difference of squares

the squares of two numbers
seperated by a minus sign.

i) $b^2 - 9 = (b + 3)(b - 3)$ ii) $16x^2 - 25y^2 = (4x - 5y)(4x + 5y)$

Note: It is important that you learn to recognize a difference of squares so that you can factor it quickly and effectively.

Examples

1. $x^2 - 25$ **Answer** $(x + 5)(x - 5)$

2. $4a^2 - 36$ $(2a - 6)(2a + 6)$

3. $b^2 - 49c^2$ $(b + 7c)(b - 7c)$

4. $9y^2 - 1$ $\hspace{3cm}$ $(3y - 1)(3y + 1)$

5. $m^2 - 81n^2$ $\hspace{3cm}$ $(m - 9n)(m + 9n)$

OPERATIONS WITH RATIONAL EXPRESSIONS

A rational expression is an algebraic expression in fractional form.

e.g. $\dfrac{2a}{b}$, $\dfrac{5}{x}$, $\dfrac{y + 6}{y - 1}$, $\dfrac{x^2 + 4x + 3}{2x + 2}$

Addition and Subtraction

i) $\dfrac{2a}{3} + \dfrac{a}{5} = \dfrac{10a}{15} + \dfrac{3a}{15} = \dfrac{10a + 3a}{15} = \dfrac{13a}{15}$

ii) $\dfrac{4}{7}x + \dfrac{y}{3} - \dfrac{17x}{21} = \dfrac{12x}{21} + \dfrac{7y}{21} - \dfrac{17x}{21} = \dfrac{12x - 17x + 7y}{21} = \dfrac{-5x + 7y}{21}$

Note. $\dfrac{4}{7}x = \dfrac{4}{7} \times \dfrac{x}{1} = \dfrac{4x}{7}$

iii) $\dfrac{1}{a} + \dfrac{2}{b} = \dfrac{b}{ab} + \dfrac{2a}{ab} = \dfrac{b + 2a}{ab}$

iv) $\dfrac{4}{x} - \dfrac{3}{5x} + \dfrac{1}{2x} = \dfrac{40}{10x} - \dfrac{6}{10x} + \dfrac{5}{10x} = \dfrac{39}{10x}$

Note: Distinguish clearly between $\dfrac{2}{3}x$ and $\dfrac{2}{3x}$

In $\dfrac{2}{3}x$, the variable, x, is associated with the numerator;

$\dfrac{2}{3}x = \dfrac{2x}{3}$

In $\dfrac{2}{3x}$, the variable, x, is associated with the denominator

Examples

1. $\dfrac{3}{4}a + \dfrac{b}{3} - \dfrac{5}{6}a$ $\hspace{2cm}$ **Answer** $\dfrac{-a + 4b}{12}$

2. $\dfrac{2}{m} - \dfrac{5}{n}$ $\hspace{3.5cm}$ $\dfrac{2n - 5m}{mn}$

3. $\dfrac{5}{3c} + \dfrac{1}{2c} - \dfrac{7}{6c}$ $\hspace{2.5cm}$ $\dfrac{1}{c}$

REDUCING TO LOWEST TERMS

In working with simple fractions earlier recall that a result such as $\frac{6}{8}$ was reduced to lowest (simplest) terms by using the principle of equivalent fractions.

e.g. $\frac{6}{8} = \frac{6 \div 2}{8 \div 2} = \frac{3}{4}$ numerator and denominator were divided by the same number (2)

or $\frac{6}{8} = \frac{3 \times \overset{1}{2}}{4 \times \underset{1}{2}} = \frac{3}{4}$ numerator and denominator were factored and the common factor (2) was divided out.

The same principle will be used to reduce rational expressions to lowest terms.

e.g. $\frac{15a}{10b} = \frac{15a \div 5}{10b \div 5} = \frac{3a}{2b}$ or $\frac{15a}{10b} = \frac{3a \times \overset{1}{5}}{2b \times \underset{1}{5}} = \frac{3a}{2b}$

$\frac{12xy}{8xz} = \frac{12xy \div 4x}{8xz \div 4x} = \frac{3y}{2z}$ or $\frac{12xy}{8xz} = \frac{3y \times \overset{1}{4x}}{2z \times \underset{1}{4x}} = \frac{3y}{2z}$

In the following examples the second (factoring) method will be used since it will not be obvious what number will divide evenly into numerator and denominator in the original rational expression.

i) $\frac{2x + 4}{x + 2} = \frac{2\,\overset{1}{(x + 2)}}{\underset{1}{(x + 2)}} = 2$

ii) $\frac{10a - 20b}{15} = \frac{\overset{2}{10}\,(a - 2b)}{\underset{3}{15}} = \frac{2\,(a - 2b)}{3}$

iii) $\frac{y^2 - 9}{y + 3} = \frac{(y - 3)\,\overset{1}{(y + 3)}}{\underset{1}{(y + 3)}} = y - 3$

iv) $\frac{a^2b - a}{ab} = \frac{\overset{1}{a}\,(ab - 1)}{\underset{1}{ab}} = \frac{ab - 1}{b}$

v) $\frac{x^2 + 7x + 6}{2x + 2} = \frac{(x + 6)\,\overset{1}{(x + 1)}}{2\,\underset{1}{(x + 1)}} = \frac{x + 6}{2}$

vi) $\frac{y^2 - 7y + 10}{y^2 - 4} = \frac{(y - 5)\,\overset{1}{(y - 2)}}{(y + 2)\,\underset{1}{(y - 2)}} = \frac{y - 5}{y + 2}$

Examples

1. $\dfrac{3x-9}{12}$ **Answer** $\dfrac{x-3}{4}$ 3. $\dfrac{y^2+6y+8}{y^2-16}$ **Ans.** $\dfrac{y+2}{y-4}$

2. $\dfrac{4a^2-8ab}{24a}$ $\dfrac{a-2b}{6}$ 4. $\dfrac{x^2-2x-3}{5x+5}$ $\dfrac{x-3}{5}$

MULTIPLICATION AND DIVISION OF RATIONAL EXPRESSIONS

Recall the method of multiplying and dividing fractions

e.g. $\dfrac{\overset{1}{\cancel{4}}}{\underset{1}{\cancel{7}}} \times \dfrac{\overset{3}{\cancel{21}}}{\underset{2}{\cancel{8}}} = \dfrac{3}{2}$ 　　$\dfrac{2}{9} \div \dfrac{5}{3} = \dfrac{2}{\underset{3}{\cancel{9}}} \times \dfrac{\overset{1}{\cancel{3}}}{5} = \dfrac{2}{15}$

If common factors could be found in numerator and denominator, then they were divided out. The same procedure will be used in operating on rational expressions.

i) $\dfrac{10}{3x-3} \times \dfrac{x-1}{15} = \dfrac{\overset{2}{\cancel{10}}}{3\underset{1}{\cancel{(x-1)}}} \times \dfrac{\cancel{(x-1)}^{1}}{\underset{3}{\cancel{15}}} = \dfrac{2}{9}$

ii) $\dfrac{y^2-36}{4x} \times \dfrac{8x}{2y+12} = \dfrac{(y-6)(y+6)}{\underset{1}{\cancel{4x}}} \times \dfrac{\overset{2\cdot1}{\cancel{8x}}}{\underset{1}{2}\underset{1}{(y+6)}} = y-6$

iii) $\dfrac{x^2+4x+3}{x^2-1} \times \dfrac{x+1}{x^2-9} = \dfrac{\cancel{(x+1)}^{1}(x+3)^{1}}{\cancel{(x+1)}_{1}(x-1)} \times \dfrac{x+1}{(x-3)\cancel{(x+3)}_{1}}$

$= \dfrac{x+1}{(x-1)(x-3)}$

iv) $\dfrac{a^2-4}{6} \div \dfrac{a^2+5a+6}{8} = \dfrac{a^2-4}{6} \times \dfrac{8}{a^2+5a+6}$

$= \dfrac{\cancel{(a+2)}^{1}(a-2)}{\underset{3}{\cancel{6}}} \times \dfrac{\overset{4}{\cancel{8}}}{\cancel{(a+2)}_{1}(a+3)} = \dfrac{4(a-2)}{3(a+3)}$

Examples

1. $\dfrac{b^2+4b-5}{12} \times \dfrac{18}{2b+10}$ **Answer** $\dfrac{3(b-1)}{4}$

2. $\dfrac{m^2-16}{m+1} \times \dfrac{m^2+2m+1}{3m-12}$ $\dfrac{(m+4)(m+1)}{3}$

3. $\dfrac{a^2+3ab}{9} \div \dfrac{2a+6b}{18}$ a

EQUATIONS AND INEQUATIONS

EQUATIONS

An equation is a statement of equality of two numbers. There are three basic parts to any equation; the left hand number (L.H.N.), the equal sign, and the right hand number (R.H.N.). Any equation can be seen as a number balance with the point of balance being the equal sign. In order not to upset this balance, you must treat the L.H.N. and R.H.N. in exactly the same manner.

SOLVING EQUATIONS

I Inspection

Simple equations may be solved by inspection (intelligent guessing).

e.g. $x + 4 = 7, x \in N$ $\qquad$ $5a = 35, a \in I$

$\Longleftrightarrow$ $x = 3$ $\qquad\qquad$ $\Longleftrightarrow$ $a = 7$

∴ the solution set is $\{3\}$ $\qquad$ ∴ the solution set is $\{7\}$

However this method is time consuming and thus inefficient for solving more complex equations.

II Formal Method

Since an equation is a number balance, then we may operate on the L.H.N. and the R.H.N. in a similar way and not destroy this balance.

RULES FOR OPERATING ON EQUATIONS

1. Addition Rule for Equations (A.R.E.)

$x - 4 = -2, x \in I$ $\qquad$ The same number (4) was added to the

$\Longleftrightarrow x - 4 + 4 = -2 + 4$ $\qquad$ L.H.N. and the R.H.N. to obtain an

$\Longleftrightarrow$ $x = 2$ $\qquad$ equivalent but simpler equation

∴ the solution set is $\{2\}$

Note: Equations are equivalent if they have the same solution set. They may be different in appearance. In solving equations, your task is to obtain simpler equivalent equations until you can state the solution set.

2. Subtraction Rule for Equations (S.R.E.)

$a + 5 = 12, a \epsilon N$ The same number (5) was subtracted

$\Longleftrightarrow a + 5 - 5 = 12 - 5$ from the L.H.N. and the R.H.N. to

$\Longleftrightarrow \quad a = 7$ obtain an equivalent but simpler

$\therefore$ the solution set is $\{7\}$ equation

3. Multiplication Rule for Equations (M.R.E.)

$\frac{1}{2}y = -4, y \epsilon I$ The L.H.N. and the R.H.N.

    were both multiplied by the

$\Longleftrightarrow (2) (\frac{1}{2}y) = (-4) (2)$ same number (2) to obtain an

    equivalent but simpler equation

$\Longleftrightarrow \quad y = -8$

$\therefore$ the solution set is $\{-8\}$

4. Division Rule for Equations (D.R.E.)

$3b = 15, b \epsilon Q$ The L.H.N. and the R.H.N.

$\Longleftrightarrow \frac{3b}{3} = \frac{-15}{3}$ were both divided by the same

    number (3) to obtain an equivalent

$\Longleftrightarrow b = -5$ but simpler equation

$\therefore$ the solution set is $\{-5\}$

In solving most equations, you will have to use a combination of these rules.

e.g. i) $3x - 8 = 1, x \epsilon N$

$\Longleftrightarrow 3x - 8 + 8 = 1 + 8$ (ARE)

$\Longleftrightarrow \quad 3x = 9$

$\Longleftrightarrow \quad \frac{3x}{3} = \frac{9}{3}$ (DRE)

$\Longleftrightarrow \quad x = 3$

$\therefore$ the solution set is $\{3\}$

ii) $5a + 6 = 3a - 4, a \epsilon I$

$\Longleftrightarrow 5a - 3a + 6 = 3a - 3a - 4$ (SRE)

$\Longleftrightarrow \quad 2a + 6 = -4$

$\Longleftrightarrow 2a + 6 - 6 = -4 - 6$ (SRE)

$\Longleftrightarrow \quad 2a = -10$

$\Longleftrightarrow \quad \frac{2a}{2} = \frac{-10}{2}$ (DRE)

$\Longleftrightarrow \quad a = -5$

$\therefore$ the solution set is $\{-5\}$

iii) $\frac{2}{3}y - 1 = 5, y \in Q$

⟺ $\frac{2}{3}y - 1 + 1 = 5 + 1$ (ARE)

⟺ $\frac{2}{3}y = 6$

⟺ $(\frac{3}{2})(\frac{2}{3}y) = (6)(\frac{3}{2})$ (MRE)

⟺ $y = 9$

∴ the solution set is $\{9\}$

Note: As demonstrated in each of these examples, the first step is to isolate the variable on one side of the equation (usually the L.H.N.) by using A.R.E. or S.R.E., then reduce the numerical coefficient of the variable to one by employing M.R.E or D.R.E.

iv) $2(a + 3) + 5 = 9, a \in R$ **Note:** If any expansion

⟺ $2a + 6 + 5 = 9$ (removal of brackets)

⟺ $2a + 11 = 9$ is necessary, then

⟺ $2a + 11 - 11 = 9 - 11$ (SRE)   it should be done

⟺ $2a = -2$ first and any like

⟺ $\frac{2a}{2} = \frac{-2}{2}$ (DRE)   terms collected

⟺ $a = -1$ before proceeding.

∴ the solution set is $\{-1\}$

v) $3(2b - 1) + 5 = 7(b + 1), b \in Q$

⟺ $6b - 3 + 5 = 7b + 7$

⟺ $6b + 2 = 7b + 7$

⟺ $6b - 7b + 2 - 2 = 7b - 7b + 7 - 2$ (SRE)

⟺ $-b = 5$ Recall $-b = (-1)b$

⟺ $(-1)(-b) = (5)(-1)$ (MRE)

⟺ $b = -5$

∴ the solution set is $\{-5\}$

vi) $\frac{2}{3}x - \frac{1}{4} = \frac{1}{2}x + \frac{3}{4}, x \in R$

⟺ $12(\frac{2}{3}x - \frac{1}{4}) = (\frac{1}{2}x + \frac{3}{4})12$ (MRE)

⟺ $12 \times \frac{2}{3}x - 12 \times \frac{1}{4} = 12 \times \frac{1}{2}x + 12 \times \frac{3}{4}$

⟺ $8x - 3 = 6x + 9$

⟺ $8x - 6x - 3 + 3 = 6x - 6x + 9 + 3$ (ARE, SRE)

⟺ $2x = 12$

⟺ $(\frac{1}{2})(2x) = (12)(\frac{1}{2})$ (MRE)

⟺ $x = 6$

∴ the solution set is $\{6\}$

Note: This example is difficult to deal with because of the fractional form of the numerical coefficients and numbers involved. To alleviate this problem, L.H.N and R.H.N. were multiplied by the lowest common denominator of the fractions (12)

A complete solution of an equation should include a verification. The following example will show the form such a solution should take.

vii) $5 - (y + 2) = 7 - 3y, \; y \in Q$

⟺ $5 - y - 2 = 7 - 3y$

⟺ $3 - y = 7 - 3y$

⟺ $3 - 3 - y + 3y = 7 - 3 - 3y + 3y$

            $2y = 4$

       $(\frac{1}{2})(2y) = (4)(\frac{1}{2})$

            $y = 2$

Verification

$LHN = 5 - (y + 2)$ $RHN = 7 - 3y$

       $= 5 - (2 + 2)$ $= 7 - 3 \cdot 2$

       $= 5 - 4$ $= 7 - 6$

       $= 1$ $= 1$

∴ $LHN = RHN$

∴ the root of the equation is 2

Note: The verification is a check of your solution. The check must be made in the original equation. If after substituting your solution into L.H.N. and R.H.N., the results are equal, then your solution has been verified.

INEQUATIONS

An inequation is a number inbalance. It is composed of a left hand number (LHN), inequality sign (>, <, $\geq$ etc) and a right hand number (RHN)

SOLVING INEQUATIONS

Rules for Operating on Inequations

The rules for operating on inequations are almost the same as those previously developed for dealing with equations save for two exceptions.

1. Addition Rule for Inequations (A.R.I.)

$$x - 5 > 4, x \in N$$
$$\Longleftrightarrow x - 5 + 5 > 4 + 5 \qquad \text{(ARI)}$$
$$\Longleftrightarrow \qquad x > 9$$

∴ the solution set
is $\{ x \mid x > 9, x \in N \}$

The same number (5) was added to the L.H.N. and the R.H.N. to obtain an equivalent but simpler inequation.

2. Subtraction Rule for Inequations (S.R.I.)

$$y + 2 < 5, y \in Q$$
$$\Longleftrightarrow y + 2 - 2 < 5 - 2 \qquad \text{(SRI)}$$
$$\Longleftrightarrow \qquad y < 3$$

∴ the solution set
is $\{ y \mid y < 3, y \in Q \}$

The same number (2) was subtracted from the L.H.N and the R.H.N. to obtain an equivalent but simpler inequation.

The addition and subtraction rules for inequations are identical to those for equations.

3. Multiplication Rule for Inequations (M.R.I.)

$$\frac{1}{3}a \geq 4, a \in I$$

$$\Longleftrightarrow (3)\left(\frac{1}{3}a\right) \geq (4)(3) \qquad \text{(MRI)}$$

$$\Longleftrightarrow \qquad a \geq 12$$

∴ the solution set is
$\{ a \mid a \geq 12, a \in I \}$

The L.H.N. and R.H.N. were multiplied by the same number to obtain an equivalent but simpler inequation.

but
$$-\frac{1}{4}y < 2, y \in Q$$

$$\Longleftrightarrow (-4)\left(-\frac{1}{4}y\right) > (2)(-4)$$

$$\Longleftrightarrow \qquad y > -8$$

∴ the solution set is
$\{ y \mid y > -8, y \in Q \}$

This is the first exception. When L.H.N. and R.H.N. are *multiplied by a negative number, the sign of inequality must be reversed.*

4. Division Rule for Inequations (D.R.I.)

$$5a \leq 10, a \in N$$

$\Longleftrightarrow \dfrac{5a}{5} \leq \dfrac{10}{5}$ (DRI)

$\Longleftrightarrow a \leq 2$

The L.H.N. and R.H.N. were divided by the same number (5) to obtain an equivalent but simpler inequation.

∴ the solution set is
$\{ a \mid a \leq 2, a \in N \}$

but $-8b \geq 16, b \in Q$

$\Longleftrightarrow \dfrac{-8b}{-8} \leq \dfrac{16}{-8}$ (DRI)

$\Longleftrightarrow b \leq -2$

This is the second exception. When L.H.N and R.H.N. are *divided by a negative number, the sign of inequality must be reversed.*

∴ the solution set is
$\{ b \mid b \leq -2, b \in Q \}$

e.g.

i)

$$2x + 5 \geq x + 7, x \in Q$$

$\Longleftrightarrow 2x - x + 5 - 5 \geq x - x + 7 - 5$

$$x \geq 2$$

∴ the solution set is $\{ x \mid x \geq 2, x \in Q \}$

The method of solving inequations is identical to that of solving equations but for the two exceptions noted in MRI and DRI.

ii)

$$y - 8 < 3y + 2, y \in Q$$

$\Longleftrightarrow y - 3y - 8 + 8 < 3y - 3y + 2 + 8$

$\Longleftrightarrow -2y < 10$

$\Longleftrightarrow \dfrac{-2y}{-2} > \dfrac{10}{-2}$

$\Longleftrightarrow y > -5$

∴ the solution set is $\{ y \mid y > -5, y \in Q \}$

See your textbook for further examples.

GEOMETRY

Geometry is a structure composed of undefined terms (point, plane, line), postulates (accepted but unproven statements) and theorems (proven statements). In this section, some of the major theorems and postulates, and their applications will be listed.

SUPPLEMENTARY ANGLES

Supplementary angles are any pair of angles which sum to a straight angle (180°)

COMPLEMENTARY ANGLES

Complementary angles are any pair of angles which sum to a right angle (90°)

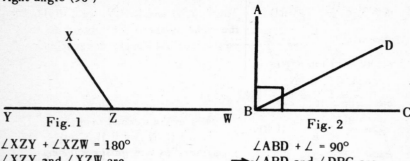

∠XZY + ∠XZW = 180°
⟹ ∠XZY and ∠XZW are supplementary angles

∠ABD + ∠ = 90°
⟹ ∠ABD and ∠DBC are complementary angles

OPPOSITE ANGLE THEOREM

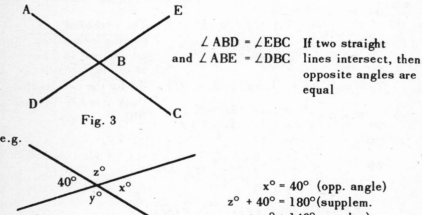

Fig. 3

∠ABD = ∠EBC   If two straight
and ∠ABE = ∠DBC   lines intersect, then opposite angles are equal

e.g.

Fig. 4

$x° = 40°$ (opp. angle)
$z° + 40° = 180°$(supplem.
∴ $z° = 140°$ angles)

$y° = z°$ (opp. angles)
∴ $y° = 140°$

ISOSCELES TRIANGLE THEOREM

Fig. 5

AB = AC   ⟺   ∠ABC = ∠ACB

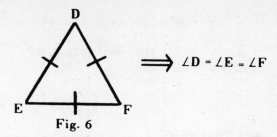

Fig. 6

$\Longrightarrow \angle D = \angle E = \angle F$

ANGLE SUM TRIANGLE THEOREM

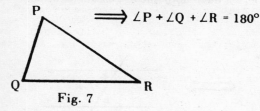

$\Longrightarrow \angle P + \angle Q + \angle R = 180°$

The sum of the interior angles of any triangle is 180°.

Fig. 7

EXTERIOR ANGLE THEOREM

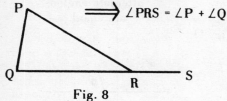

$\Longrightarrow \angle PRS = \angle P + \angle Q$

The exterior angle of any triangle equals the sum of the two interior and opposite angles.

Fig. 8

e.g.   i)

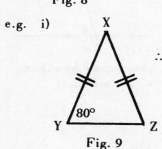

Fig. 9

$\angle Z = 80°$ (isosc $\triangle$ th^{m})
$\angle X + \angle Y + \angle Z = 180°$ (angle sum $\triangle$ th^{m})
$\therefore \ \angle X + 80° + 80° = 180°$
$\angle X = 20°$

ii)

Fig. 10

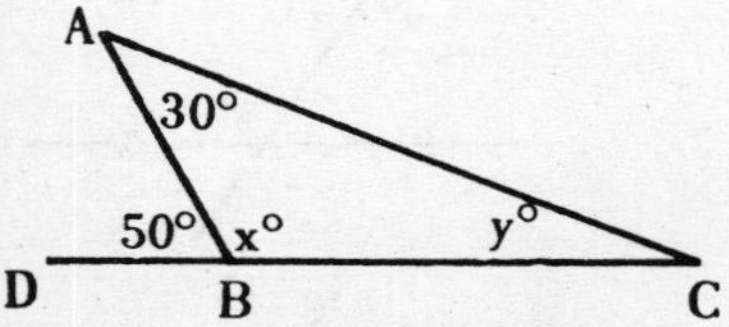

$50° = y° + 30°$ (ext angle th^{m})
$\therefore \ y° = 20°$
$x° + y° + 30° = 180°$
$x° + 20° + 30° = 180°$
$\therefore \ x° = 130°$

or

52

$$x° + 50° = 180° \text{ (straight angle)}$$
$$x° = 130°$$
$$x° + y° + 30° = 180° \text{ (angle sum } \Delta \text{ th}^m\text{)}$$
$$130° + y° + 30° = 180°$$
$$\therefore y° = 20°$$

PARALLEL LINE THEOREMS

Parallel lines are lines which have no point in common (i.e. do not intersect)

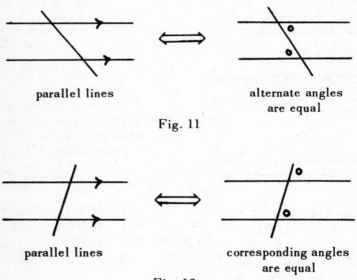

parallel lines

alternate angles are equal

Fig. 11

parallel lines

corresponding angles are equal

Fig. 12

e.g. i)
$$x° = 140° \text{ (corresp. angles)}$$
$$y° = x° \text{ (altern. angles)}$$
$$\therefore y° = 140°$$
$$z° + 140° = 180°$$
$$\therefore z° = 40°$$

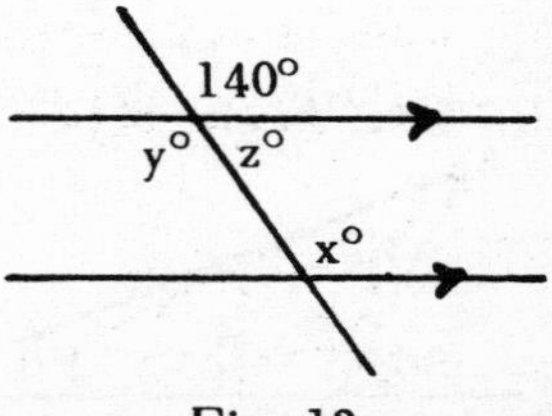

Fig. 13

ii)

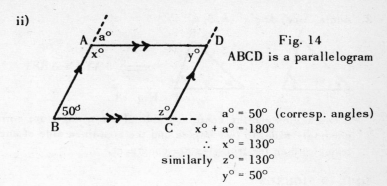

Fig. 14
ABCD is a parallelogram

$a^\circ = 50^\circ$ (corresp. angles)
$x^\circ + a^\circ = 180^\circ$
$\therefore \quad x^\circ = 130^\circ$
similarly $z^\circ = 130^\circ$
$y^\circ = 50^\circ$

CONGRUENCY POSTULATES

Two figures are congruent if they are equal in all respects (corresponding sides equal, corresponding angles equal and areas equal).

If is very useful in dealing with congruent triangles that care be taken to match up corresponding vertices so that corresponding sides and angles can be easily established.

e.g

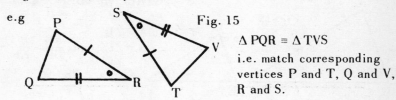

Fig. 15

$\Delta \, PQR \equiv \Delta \, TVS$

i.e. match corresponding vertices P and T, Q and V, R and S.

1. Side, Side, Side (S, S, S)

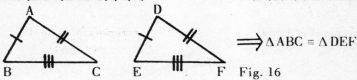

$\Longrightarrow \Delta \, ABC \equiv \Delta \, DEF$

Fig. 16

If one triangle has its three sides correspondingly equal to three sides of another triangle, then the triangles are congruent.

2. Side, Angle, Side (S, A, S)

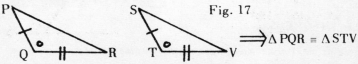

Fig. 17

$\Longrightarrow \Delta \, PQR \equiv \Delta \, STV$

If two sides and the *contained angle* of one triangle are correspondingly equal to two sides and the contained angle of another triangle, then the triangles are congruent.

3. Angle, Side, Angle (A, S, A)

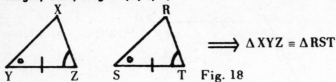

$$\Longrightarrow \triangle XYZ \equiv \triangle RST$$

Fig. 18

If two angles and the contained side of one triangle are correspondingly equal to two angles and the contained side of another triangle, then the triangles are congruent.

SIMILAR FIGURES

Whereas congruent figures are equal in all respects (sides, angles and area), similar figures have the same shape (equal angles) but corresponding sides are not equal. However there is a special relationship amongst the corresponding sides; they are in equal ratio (i.e. corresponding sides when compared by division yield the same result)

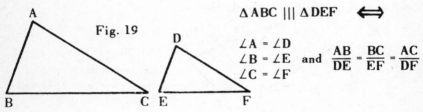

Fig. 19

$$\triangle ABC \;|||\; \triangle DEF \iff$$

$$\angle A = \angle D$$
$$\angle B = \angle E \quad \text{and} \quad \frac{AB}{DE} = \frac{BC}{EF} = \frac{AC}{DF}$$
$$\angle C = \angle F$$

1. Angle, Angle, Angle (A, A, A)

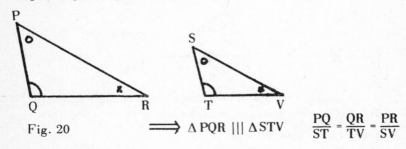

Fig. 20

$$\Longrightarrow \triangle PQR \;|||\; \triangle STV$$

$$\frac{PQ}{ST} = \frac{QR}{TV} = \frac{PR}{SV}$$

CIRCLE THEOREMS

a. Angle at the centre and angle at the circumference.

$\angle BOC$ is an angle at the centre *subtended by the arc BXC*, or an angle at the centre *standing on the arc BXC*.

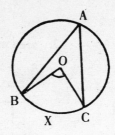

$\angle BAC$ is an angle at the circumference *subtended by the arc BXC*, or an angle at the circumference *standing on the arc BXC*.

If an angle at the centre and an angle at the circumference are subtended by a common arc, then the measure of the angle at the centre is twice the measure of the angle at the circumference.

i.e. $\angle BOC = 2\angle BAC$

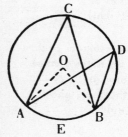

If angles at the circumference are subtended by a common arc, then the angles are equal.

i.e. $\angle ACB = \angle ADB$

The angle in a semicircle is a right angle.

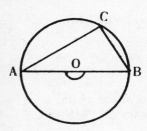

If an angle at the circumference is subtended by a semi-circle, then the measure of the angle is 90°

i.e. $\angle ACB = 90°$

AREA

1. Area of a Rectangle

$A = l \times w$ (sq. units)

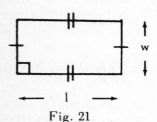

Fig. 21

2. Area of a Square

$A = a \times a$ (sq. units)
$= a^2$

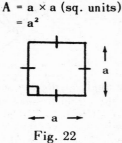

Fig. 22

3. Area of a Parallelogram

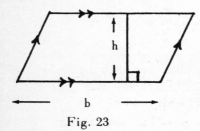

Fig. 23

$A = b \times h$ (sq. units)
$= (base) \times (height)$

The area of a parallelogram is the product of one of its sides (base) and the perpendicular distance (height) between that side and the opposite parallel side.

4. Area of a Triangle

$A = \frac{1}{2} \times b \times h$ (sq. units)

$= \frac{1}{2} \times (base) \times (height)$

Fig. 24

The area of a triangle is one half the product of one of its sides (base) and the perpendicular distance (height) between that side and the opposite vertex.

e.g. i)

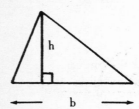

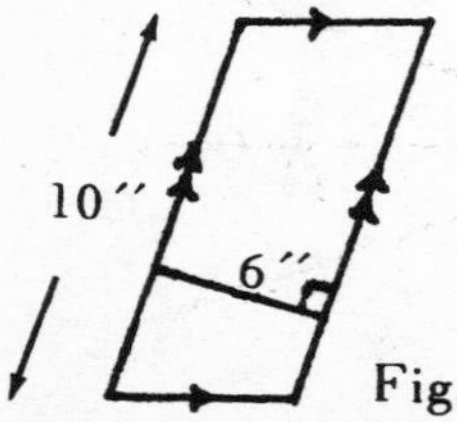

Fig. 25

$A = b \times h$
$= 10 \times 6$
$= 60$ (sq. in.)

ii)

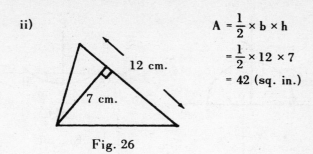

$A = \dfrac{1}{2} \times b \times h$

$= \dfrac{1}{2} \times 12 \times 7$

$= 42$ (sq. in.)

12 cm.

7 cm.

Fig. 26

GEOMETRY CONSTRUCTIONS

Constructions and their descriptions: There are six fundamental constructions which you should know. These are shown diagrammatically here.

1. To construct an angle equal to a given angle.

2. To bisect a given angle.

3. To construct a perpendicular at a point on the line.

4. To construct a perpendicular from a point NOT on the line.

5. To construct the right bisector of a given line.

6. To construct a line parallel to a given line and passing through a given point.

1.

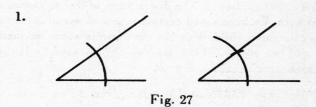

Fig. 27

2.

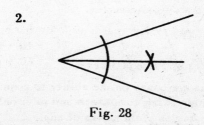

Fig. 28

58

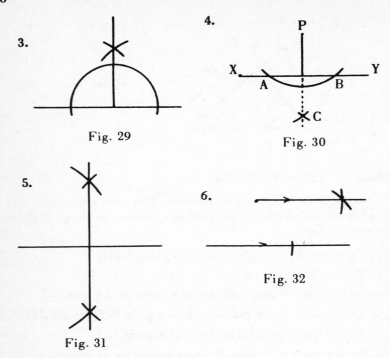

3.

Fig. 29

4.

P

X —— A B —— Y

C

Fig. 30

5.

Fig. 31

6.

Fig. 32

Describing constructions: The description of No. 4, above, is included here. We have noted certain standard words in the description notably the ones which describe the drawing of an arc, and the joining of two points. These are "stock" phrases for the mathematician.

No. 4. *With Centre P and suitable radius, describe an arc* to cut XY in A and B.
With Centres A and B *and equal radii, describe arcs* to cut at C.
Join PC
Then PC ⊥ XY.

Constructing Angles of Various Sizes:

All constructions are based on the ability to construct a 90° angle and a 60° angle using compasses and ruler only. Further bisection of these angles yield series of angles.

The 90° Constructions

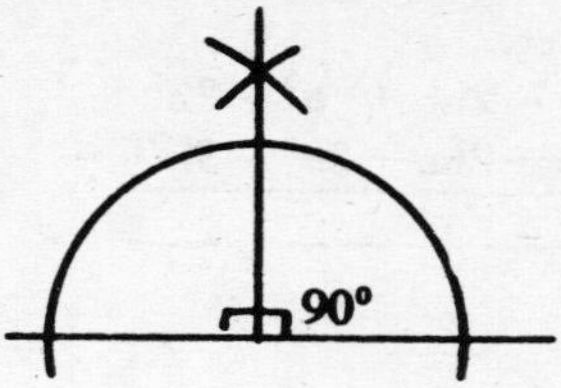

Fig. 33

The 60° Constructions

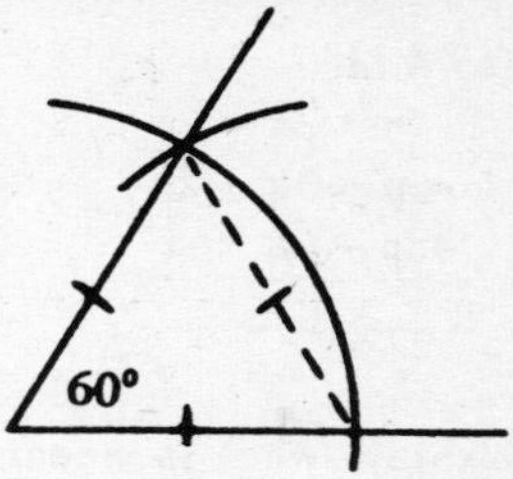

60°

Fig. 34

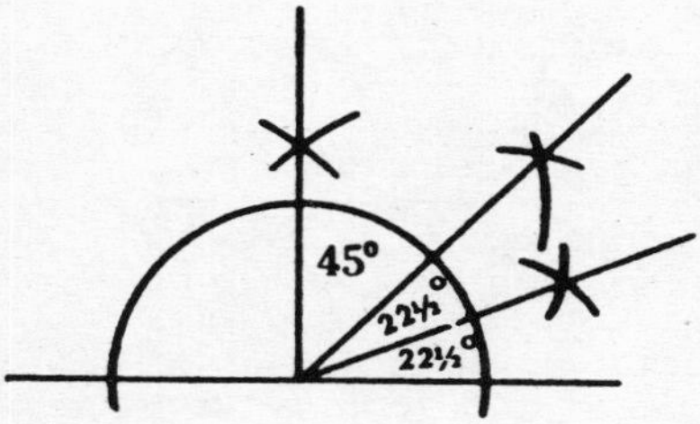

45°
22½°
22½°

Fig. 35

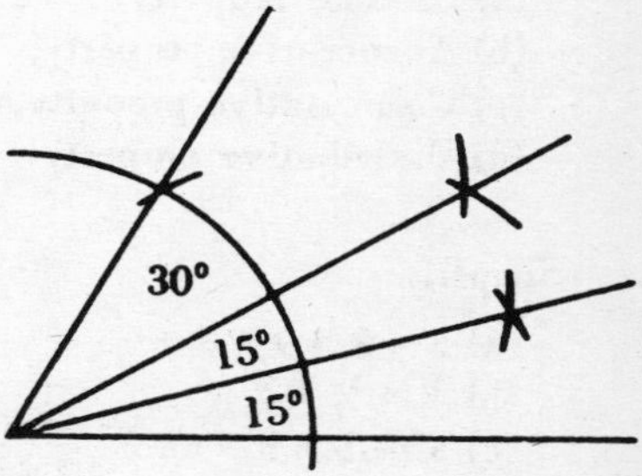

30°
15°
15°

Fig. 36

Combined Series

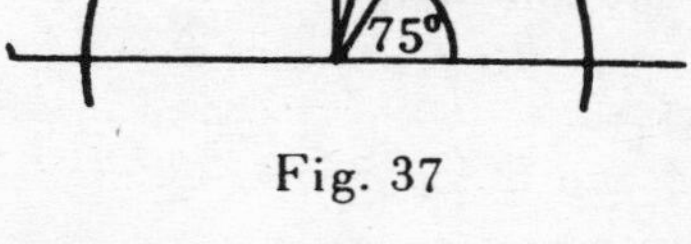

75°

Fig. 37

EXAMINATIONS AND THEIR SOLUTIONS

The following collection of suggested examinations is not intended to indicate where you should be at any given time of the year,
nor does it indicate a *standard* degree of difficulty. It should, however, provide information regarding the coverage by *some* teachers
at Christmas, and will, we hope, show a variety of question types
which should help students to grow mathematically as well as to
gain computative skills. Where domains have not been specified,
the domain R is assumed

CHRISTMAS EXAMINATION – SAMPLE 1

1. (a) Add:

    $4p + 3q - r$
    $-p - 3q + 2r$
    $-2p - q + r$
    _____

    (b) Subtract:

    $-a + 3b - 2bc + a^2 - 2b^2$
    $-a - 2b + 2bc + 2a^2 - 3b^2$
    _____

2. Explain what is meant by each of the following, and give an illustrative example in each case.

    (a) Closure property,
    (b) Associative property,
    (c) Commutative property,
    (d) Distributive property.

3. Simplify:

    (a) $5 \times 8 + 9 \div 3$
    (b) $\frac{2}{3} + \frac{3}{7} + \frac{1}{2}$
    (c) $a^2 - b \times b + aa$
    (d) $4a + 3ab - 2ab + 2a$
    (e) $(^-13)(^-5) + {}^-13 \times {}^-17$

4. Evaluate each of the following where $x = 7$, $y = 3$, $z = o$:

    (a) $3(x + y)$
    (b) $2x (y + z)$
    (c) $x y z$
    (d) $xy + yz + zx$

5. Factor each of the following:

    (a) $ab + ac$
    (b) $x^2 + 4x + 4$
    (c) $x^2 - 4$
    (d) $cx^2 - 10cx - 24c$

6. Graph the solution of each of the following on a suitable number line or segment thereof:

    (a) $\{ x \mid x + 5 = 3, x \epsilon N \}$
    (b) $\{ x \mid x + 5 = 3, x \epsilon I \}$
    (c) $\{ y \mid y - 2 > 3, y \epsilon I \}$

7. State whether each of the following is true or false
   (a) $3 \geq 2$
   (b) The answer, 25, is obtained from $3 + 8 + 14$ by the closure property.
   (c) $8 - 2 > 6$
   (d) $0 \times 3 > 4$
   (e) $2 \div 7 = 7 \div 2$

8. In each part of this question you will find two numerals. State whether or not they stand for the *same* number, in each part:
   (a) $4(a + b)$
       $4 + a \times b$
   (b) $(6a)(-2bc)(-ab^2c)$
       $(3b^3)(2a^2c^2)(-2)$
   (c) $3 \times 7 + 2 \times 8$
       $7 + 2 \times 8 + 3$
   (d) $16 \div 4 + 4$
       $16 \div (4 + 4) + 6$
   (e) $3(a + b)(2)$
       $6a + b$

9. Solve each of the following and check your answer:
   (a) $2x + 7x + 8x = 0,\ x \epsilon N_0$
   (b) $4x = 8 + 2x,\ x \epsilon I$
   (c) $12x + 35 = 7x,\ x \epsilon I$

10. The sum of three consecutive numbers is 39
    (a) Write an algebraic equation which expresses this;
    (b) Solve the equation you wrote for (a) and *name the three numbers*.
    (c) Is the sum of the numbers found 39?

This solution set, and subsequent ones, show a suggested form of answer which you could copy. Make your answer sufficiently complete that the examiner can identify the question you are answering without reference to the examination question.

62

SET OF SOLUTIONS TO CHRISTMAS EXAMINATION SAMPLE 1

1. (a) $\begin{array}{r} 4p + 3q - r \\ -p - 3q + 2r \\ -2p - q + r \\ \hline p - q + 2r \end{array}$

    (b) $\begin{array}{r} a + 3b - 2bc + a^2 - 2b^2 \\ -a - 2b + 2bc + 2a^2 - 3b^2 \\ \hline 5b - 4bc - a^2 + b^2 \end{array}$

2. (a) *Closure property* —that property of a set of numbers such that when an operation is performed on two numbers from the set, a unique answer is found, and this answer is another number of the set.

    > *Example*: 2, 3 ϵN
    > 2 + 3 = 5, and 5 ϵN
    > ∴ The set N is closed under addition.

    (b) *Associative property* — that property of an operation which permits the association, or grouping, of the numbers to be operated on in different ways without changing the answer.

    > *Example:* (a + b) + c = a + (b + c)
    > The associative property of addition is illustrated.

    (c) *Commutative property:* that property of an operation which permits the reversing of the operation without changing the answer.
    > *Example:* a x b = b x a
    > The commutative property for multiplication is illustrated.

    (d) *Distributive property:* that property of operations which shows the use of multiplication over addition.
    > *Example:* 2(x + y) = 2x + 2y

3. (a) $5 \times 8 + 9 \div 3$
    = 40 + 3
    = 43

    (b) $\frac{2}{3} + \frac{3}{7} + \frac{1}{2}$
    $= \frac{28 + 18 + 21}{42}$
    $= \frac{67}{42}$
    $= 1\frac{25}{42}$

(c) $a^2 - b \times b + aa$
     $= a^2 - b^2 + a^2$
     $= a^2 + a^2 - b^2$
     $= 2a^2 - b^2$

(d) $4a + 3ab - 2ab + 2a$
     $= 4a + 2a + 3ab - 2ab$
     $= 6a + ab$

(e) $(-13)(-15) + -13 \times -17$
     $= +65 + +221$
     $= +286$

4.   (a) $x = 7$, $y = 3$, $z = 0$
        $3(x + y)$
       $= 3(7 + 3)$
       $= 3(10)$
       $= 30$

   (b) $2x(y + z)$
       $= 14(3 + 0)$
       $= 14(3)$
       $= 42$

   (c) xyz
       $= (7)(3)(0)$
       $= 0$

   (d) $xy + yz + zx$
       $= (7)(3) + (3)(0) + (0)(7)$
      $= 21$

5.   (a) $ab + ac$
      $= a(b + c)$

   (b) $x^2 + 4x + 4$
      $= (x + 2)^2$

   (c) $x^2 - 4$
      $= (x + 2)(x - 2)$

   (d) $cx^2 - 10cx - 24c$
      $= c(x^2 - 10x - 24)$
      $= c(x - 12)(x + 2)$

6.   (a) $\{x \mid x + 5 = 3, x \epsilon N\}$
         $x + 5 = 3$
     ⬤➤$x = -2$

        The solution set is ϕ because $-2 \notin N$

  (b) $\{x \mid x + 5 = 3, x \epsilon I\}$
         $x + 5 = 3$
      ⬤➤$x = -2$

        $^-3 \ ^-2 \ ^-1 \ 0 \ ^+1 \ ^+2 \ ^+3 \ ^+4$ I-line

  (c) $\{y \mid y - 2 > 3 \ y \epsilon I\}$
         $y - 2 > 3$
      ⬤➤$y > 5$

        $^+3 \ ^+4 \ ^+5 \ ^+6 \ ^+7 \ ^+8 \ ^+9 \ ^+10$ I-line

7.   (a) $3 \geq 2$ (false)

  (b) The answer, 25, is obtained from $3 + 8 + 14$ by the closure property. (false)
      (Note that this is an illustration of the closure property, but the result is found from the use of the associative property for addition)

  (c) $8 - 2 > 6$
     ⬤➤$6 > 6$ (false)

  (d) $0 \times 3 > 4$
     ⬤➤$0 > 4$ (false)

  (e) $2 \div 7 = 7 \div 2$ (false) Division is not commutative

8.   (a) $4(a + b) = 4a + 4b$
       $4 + a \times b = 4 + ab$
      The given numerals do not stand for the same number.

  (b) $(6a)(-2bc)(-ab^2c) = 12a^2b^3c^2$
      $(3b^3)(2a^2c^2)(-2) \quad = -12a^2b^3c^2$
      The given numerals do not stand for the same number.

  (c) $3 \times 7 + 2 \times 8 = 21 + 16 = 37$
      $7 + 2 \times 8 + 3 = 7 + 16 + 3 = 26$
      The given numerals do not stand for the same number.

(d) $16 \div 4 + 4 = 4 + 4 = 8$

$16 \div (4 + 4) + 6 = 2 + 6 = 8$

The given numerals stand for the same number.

(e) $3(a + b)(2) = 6a + 6b$

$6a + b$

The given numerals do not stand for the same number.

9. **(a)** $2x + 7x + 8x = 0, \ x \ \epsilon \ \text{No}$

$\Leftrightarrow \qquad 17x = 0$

$\Leftrightarrow \qquad \dfrac{17x}{17} = \dfrac{0}{17}$

$\Leftrightarrow \qquad x = 0$

Check LHN $= 2x + 7x + 8x$ \qquad RHN $= 0$

$= 2 \times 0 + 7 \times 0 + 8 \times 0$

$= 0 + 0 + 0$

$= 0$

$\therefore \ 0$ is the root of the equation.

(b) $\qquad 4x = 8 + 2x, \ x \ \epsilon \ I$

$\Leftrightarrow 4x - 2x = 8 + 2x - 2x$

$\Leftrightarrow \qquad 2x = 8$

$\Leftrightarrow \qquad x = 4$

Check LHN $= 4x$ \qquad\qquad RHN $= 8 + 2x$

$= 4 \times 4$ \qquad\qquad\qquad $= 8 + 2 \times 4$

$= 16$ \qquad\qquad\qquad\qquad $= 16$

$.: \ 4$ is the root of the equation.

(c) $\qquad\qquad 12x + 35 = 7x, \ x \ \epsilon \ I$

$\Leftrightarrow 12x - 7x + 35 - 35 = 7x - 7x - 35$

$\Leftrightarrow \qquad\qquad 5x = -35$

$\Leftrightarrow \qquad\qquad \dfrac{5x}{5} = \dfrac{-35}{5}$

$\Leftrightarrow \qquad\qquad x = -7$

Check LHN $= 12x + 35$ \qquad RHN $= 7x$

$= 12 \ (-7) + 35$ \qquad\qquad $= 7 \ (-7)$

$= -84 + 35$ \qquad\qquad\qquad $= -49$

$= -49$

$\therefore$ the root of the equation is -7.

10. (a) Let three consecutive numbers be x, x+ 1, x + 2
then x + x + 1 + x + 2 = 39

   (b) x + x + 1 + x + 2 = 39
$$3x + 3 = 39$$
$$3x = 36$$
$$x = 12$$

   The required numbers are 12, 13, 14

   (c) 12 + 13 + 14 = 39 √

CHRISTMAS EXAMINATION SAMPLE – 2
(time suggested – 1 1/2 hr.)

1. (a) Add:

$$-m + n + p$$
$$m - n - p$$
$$m - n + p$$

   (b) Subtract:

$$m^2 + n^2 + p^2 - x^2 - y^2$$
$$-m^2 + n^2 + 2p^2 + x^2 - y^2$$

2. Name the property illustrated by each of the following:

   (a) $(3 + 5) + 9 = 3 + (5 + 9)$

   (b) $3(x + y) = 3x + 3y$

   (c) $5 + 18 = 23$ and $23 \in N$

   (d) $2 \times 3 = 3 \times 2$

3. Simplify:

   (a) $2x + 3xy - x - 5xy$

   (b) $(5a - 3b) + (2a + b) + (-a - b)$

   (c) $(6x + 2y) - (x + 2y)$

4. Simplify:

   (a) $(2)(-3)(-1) + (21) \div (-3)$

   (b) $(-1)^3 + (-2)^2$

   (c) $-\dfrac{2}{3} + \dfrac{5}{6} - 1\dfrac{1}{2}$

5. By factoring the numerator and denominator of each fraction and dividing out like factors reduce each to lowest terms:

   (a) $\dfrac{3a + 6b}{ax^2 + 2bx^2}$

(b) $\dfrac{(c + d)x}{c^2 - d^2}$

(c) $\dfrac{ad + bd + cd}{dx}$

6. Graph the solution set on a suitable number line or segment thereof:

   (a) $\{ a \mid 3a + 2 = a - 1, a \in I \}$

7. State whether each of the following is true or false

   (a) The set N is a subset of I

   (b) $\frac{1}{7}$ is a rational number because its decimal form shows it to be a recurring decimal fraction.

   (c) the number one is not a prime number

   (d) a member of the solution set of the inequality $3x - 2 \geq 7$ is 5

   (e) division by zero is impossible because the answer is undefined

8. Equivalent equations and inequalities have the same solution by definition). Which of the following pairs of equations or inequalities are "equivalent pairs"?

   (a) $5a > 2$, $(-5) a > (-2)$

   (b) $3p > 2 + p$, $2p > 2$

   (c) $r < -2$, $-r < 2$

   (d) $5 - r < r - 1$, $2r < 6$

9. Solve each of the following and check your answer:

   (a) $6x + 2 = 4x - 8$

   (b) $6(x - 7) = 10(x - 7)$

   (c) $3(x + 3) - 5 = 3(x + 1) + 1$

10. Mrs. Brown purchased 3x packets of butter, 5x tins of fruit juice and 2x boxes of kleenex. Altogether she took home 30 separate articles.

    (a) Write an algebraic equation showing the above

    (b) Solve the equation you wrote in (a) and state how many articles of each kind there are.

SET OF SOLUTIONS TO CHRISTMAS EXAMINATION – SAMPLE 2

1. (a)
$$-m + n + p$$
$$m - n - p$$
$$m - n + p$$
$$\overline{m - n + p}$$

(b)
$$m^2 + n^2 + p^2 - x^2 - y^2$$
$$-m^2 + n^2 + 2p^2 + x^2 - y^2$$
$$\overline{2m^2 \qquad - p^2 - 2x^2}$$

2. (a) $(3 + 5) + 9 = 3 + (5 + 9)$ **Associative property**

(b) $3(x + y) = 3x + 3y$ **Distributive property**

(c) $5 + 18 = 23$ and $23 \epsilon N$ **Closure property**

(d) $2 \times 3 = 3 \times 2$ **Commutative property**

3. (a) $2x + 3xy - x - 5xy$
$$= x - 2xy$$

(b) $(5a - 3b) + (2a + b) + (-a - b)$
$$= 5a - 3b + 2a + b - a - b$$
$$= 6a - 3b$$

(c) $(6x + 2y) - (x + 2y)$
$$= 6x + 2y - x - 2y$$
$$= 5x$$

4. (a) $(2)(-3)(-1) + (21) \div (-3)$
$$= -6 + (-7)$$
$$= -13$$

(b) $(-1)^3 + (-2)^2$
$$= (-1)(-1)(-1) + (-2)(-2)$$
$$= -1 + (+4)$$
$$= 3$$

(c) $-\dfrac{2}{3} + \dfrac{5}{6} - 1\dfrac{1}{2}$
$$= \dfrac{-4}{6} + \dfrac{5}{6} - \dfrac{9}{6}$$
$$= \dfrac{-8}{6} = -\dfrac{4}{3}$$

5. (a) $\dfrac{3a + 6b}{ax^2 + 2bx^2}$
$$= \dfrac{3\overset{1}{\cancel{(a + 2b)}}}{x^2 \underset{1}{\cancel{(a + 2b)}}}$$
$$= \dfrac{3}{x^2}$$

(b) $\dfrac{(c + d) x}{c^2 - d^2}$

$= \dfrac{(c + d) x}{(c + d)(c - d)}$

$= \dfrac{x}{c - d}$

(c) $\dfrac{ad + bd + cd}{dx}$

$\dfrac{d(a + b + c)}{dx}$

$= \dfrac{a + b + c}{x}$

6. (a) $\{ a \mid 3a + 2 = a - 1, a \in I \}$

$3a + 2 = a - 1$

⟺ $3a - a = 1$

⟺ $2a = 1$

⟺ $a = \frac{1}{2}$

The solution set is ϕ because $\frac{1}{2} \notin I$

No graph of this can be drawn

7. (a) false (note that the members of N are signless)

(b) false

(c) true

(d) true

(e) true

8. (a) $5a > 2$ $(-5) a > (-2)$

   $a > \frac{2}{5}$ $5a < 2$

   $a < \frac{2}{5}$

These inequalities are NOT equivalent.

(b) $3p > 2 + p$ $2p > 2$

   $2p > 2$ $p > 1$

   $p > 1$

These inequalities are equivalent.

(c) $r < -2$ $r < 2$

   $r > -2$

These inequalities are NOT equivalent

(d)
$$5 - r < r - 1 \qquad\qquad 2r < 6$$
$$-r - r < -1 -5 \qquad\qquad r < 3$$
$$-2r < -6$$
$$r > 3$$

These inequalities are NOT equivalent.

9.   **(a)**
$$6x + 2 = 4x - 8$$
$$\Longleftrightarrow 6x - 4x + 2 - 2 = 4x - 4x - 8 - 2$$
$$\Longleftrightarrow \qquad\qquad 2x = -10$$
$$\Longleftrightarrow \qquad\qquad \frac{2x}{2} = \frac{-10}{2}$$
$$\Longleftrightarrow \qquad\qquad x = -5$$

LHN = 6x + 2   RHN = 4x - 8
      = 6 (-5) + 2   = 4 (-5) - 8
      = -30 + 2 = -28   = -20 - 8 = -28

∴ -5 is the root of the equation

(b)
$$6 (x - 7) = 10 (x - 7)$$
$$\Longleftrightarrow \qquad 6x - 42 = 10x - 70$$
$$\Longleftrightarrow 6x - 10x - 42 + 42 = 10x - 10x - 70 + 42$$
$$\Longleftrightarrow \qquad\qquad -4x = -28$$
$$\Longleftrightarrow \qquad\qquad \frac{-4x}{-4} = \frac{-28}{-4}$$
$$\Longleftrightarrow \qquad\qquad x = 7$$

LHN = 6 (x - 7)   RHN = 10 (x - 7)
      = 6 (7- 7)   = 10 (7 - 7)
      = 6 (0)   = 10 (0)
      = 0   = 0

∴ 7 is the root of the equation.

(c)
$$3 (x + 3) - 5 = 3 (x + 1) + 1$$
$$\Longleftrightarrow \quad 3x + 9 - 5 = 3x + 3 + 1$$
$$\Longleftrightarrow \qquad 3x + 4 = 3x + 4$$
$$\Longleftrightarrow 3x - 3x + 4 - 4 = 3x - 3x + 4 - 4$$
$$\Longleftrightarrow \qquad\qquad 0 = 0$$

This indicates that the given equation is an identity, that is, the equation is true for all values of the variable.

10.   **(a)** 3x + 5x + 2x = 30

(b) $3x + 5x + 2x = 30$

$\Longleftrightarrow \qquad 10x = 30$

$\Longleftrightarrow \qquad x = 3$

The number of articles was $3 \times 3 = 9$ pounds of butter

$5 \times 3 = 15$ tins of fruit

$2 \times 3 = 6$ boxes of kleenex

Total 30

CHRISTMAS EXAMINATION — SAMPLE 3
(suggested time — 1 1/2 hr.)

1. (a) Add:

$2r - 3(a + b) + z$

$-5r + 5(a + b) - 2z$

$7r + (a + b) + 3z$

(b) Subtract:

$-3de - 4ac - 2xy - yz - ar$

$3de + ac - xy - 4yz + 3ar$

2. In each of the following cases insert the proper signs to give the desired result:

(a) 3   5   2 = 10

(b) 5   3   2 = 4

(c) 5   4   = 20

(d) 27   3   = 9

(e) 8   7   2   5   = 46

3. Evaluate each of the following when $a = 2$, $b = 7$, $c = 1$.

(a) $3a - (ab - c)$

(b) $\dfrac{5ab - b}{a + 2b}$

(c) $\dfrac{a^2 + 6b + 9}{a^3 - b^2}$

4. List the members of each of the following sets in an appropriate fashion:

(a) $D = \{x \mid x + 1 = 5, x\epsilon N\}$

(b) $E = \{y \mid y + 2 > 3, \ y + 2 < 8, \ y\epsilon N\}$

(c) $F = \{a \mid a > -3. \ a \leq 3, \ a\epsilon I\}$

5. **Factor each of (a) and (b):**

   (a) $a^2b + ab^2$

   (b) $a^2 + 2ab + b^2$

6. **Write an equation, or an inequality, which could have as its solution set the indicated graph. (one for each)**

   l-line

   l-line

   l-line

7. **Simplify each of the following:**

   (a) $(2b + 4c + bc) + (bc - 2c - 2b)$

   (b) $3a(2b - c) - (3ac + 2b)$

   (c) $x + (2y + z) - (x + y - z) + 2x$

   (d) $(3a + b)(2a - b) + (a - b)(2a + 5b)$

8. **In each of the following, a pair of numerals are listed. Evaluate them where necessary and state whether or not they stand for the same number.**

   (a) $3 + 5 \times 4$, 23

   (b) $3 + 4 \times 2$, $(3 + 4) \times 2$

   (c) $4 \times 3 - 2$, 10

   (d) $8 \div 4 + 3$, $(8 \div 4) + 3$

   (e) $(^5/_2 + ^4/_3) + 2$, $^5/_2 + (^4/_3 + 2)$

9. **Write an algebraic expression "translating" each of the following:**

   (a) The amount of money I would have if I had x cents and someone gave me a quarter.

   (b) The next even integer to x, if x is even.

   (c) The number of passengers on a busload of x passengers after 3 got off.

   (d) Peter's age 4 years ago if his present age is y.

   (e) A number 8 less than 5x.

10. Given the set $0 = \{-1, 2, -3, 4\}$, find the set of all negative numbers obtained as products of pairs of elements from 0.

SET OF SOLUTIONS TO CHRISTMAS EXAMINATION – SAMPLE 3

1. (a) $2r - 3(a + b) + z$
   $-5r + 5(a + b) - 2z$
   $7r + (a + b) + 3z$
   $\overline{4r + 3(a + b) + 2z}$

   (b) $-3de - 4ac - 2xy - yz - a$
   $3de + ac - xy - 4yz + 3a$
   $\overline{-6de - 5ac - xy + 3yz - 4a}$

2. (a) $3 + 5 + 2 = 10$
   (b) $(5 + 3) \div 2 = 4$
   (c) $5 \times 4 = 20$
   (d) $27 \div 3 = 9$
   (e) $8 \times 7 - 2 \times 5 = 46$

3. (a) $3a - (ab - c)$
   $= 3(2) - (14 - 1)$
   $= 6 - 13$
   $= -7$

   (b) $\dfrac{5ab - b}{a + 2b}$
   $= \dfrac{5(14) - 7}{2 + 14}$
   $= \dfrac{70 - 7}{16}$
   $= \dfrac{63}{16}$
   $= 3\dfrac{15}{16}$

   (c) $\dfrac{a^2 + 6b + 9}{a^3 - b^2}$
   $= \dfrac{(2)(2) + 6(7) + 9}{(2)(2)(2) - (7)(7)}$
   $= \dfrac{4 + 42 + 9}{8 - 49}$
   $= \dfrac{55}{-41}$
   $= -1\dfrac{14}{41}$

4. (a) $D = \{x \mid x + 1 = 5, \ x \in N\}$

$$x + 1 = 5$$
$$x = 4$$
$$D = \{4\}$$

(b) $E = \{y \mid y + 2 > 3, \ y + 2 < 8, \ y \in N\}$

$y + 2 > 3$	$y + 2 < 8$
$y > 1$	$y < 6$

$$E = \{2, 3, 4, 5, \}$$

(c) $F = \{a \mid a > -3, \ a \leq 3, \ a \in I\}$

$$a > -3 \quad , \quad a \leq 3$$
$$F = \{-2, -1, 0, +1, +2, +3\}$$

5. (a) $a^2b + ab^2$
$$= ab(a + b)$$

(b) $a^2 + 2ab + b^2$
$$= (a + b)(a + b)$$

6. (a) $x = 3$ or any equation which may be "reduced" to this

(b) $x \leq -2$ or any equivalent inequality.

(c) $x + 4 = 4x + 3$ or any equivalent equation.

7. (a) $(2b + 4c + bc) + (bc - 2c - 2b)$

$$= 2b + 4c + bc + bc - 2c - 2b$$

$$= 2c + 2bc$$

(b) $3a(2b - c) - (3ac + 2b)$

$$= 6ab - 3ac - 3ac - 2b$$

$$= 6ab - 6ac - 2b$$

(c) $x + (2y + z) - (x + y - z) + 2x$

$$= x + 2y + z - x - y + z + 2x$$

$$= 2x + y + 2z$$

(d) $(3a + b)(2a - b) + (a - b)(2a + 5b)$

$$= 6a^2 - ab - b^2 + 2a^2 + 3ab - 5b^2$$

$$= 8a^2 + 2ab - 6b^2$$

8. (a) 3 + 5 x 4
    = 3 + 20
    = 23

   ∴ The two numerals stand for the same number.

   (b) 3 + 4 x 2   (3 + 4) x2
       = 3 + 8   = 7 x 2
       = 11   = 14

   ∴ The two numerals do NOT represent the same number.

   (c) 4 x 3 – 2
       = 12 – 2
       = 10

      ∴ The two numerals represent the same number.

   (d) 8 ÷ 4 + 3 – 5

      Because both numerals are the same as far as order of operations are concerned, they stand for the same number.

   (e) $(^5\!/_2 + {}^4\!/_3) + 2$

       $= (\frac{15 + 8}{6}) + 2$

       $= 3\frac{5}{6} + 2$

       $= 5\frac{5}{6}$

      Because addition is associative, both numerals represent the same number.

9. (a) x + 25
   (b) x ± 2
   (c) x – 3
   (d) y – 4
   (e) 5x – 8

10. Given the set 0 = { –1, 2, –3, 4 }
    The set of negative products is T = { –2, –4, –6, –12 }

EASTER EXAMINATION – SAMPLE 1
(suggested time – 2 hr.)

1. Define the set of numbers represented by each of N_0, Q, I.

2. Equations or inequalities and number lines have been drawn in pairs. Examine each pair, and state whether or not the graphs represent the solution sets of the corresponding equations.

    (a) $\{x \mid x - 3 < 2, x \in N\}$

    (b) $\{x \mid x + 3 = 8, x \in N\}$

    (c) $\{a \mid 5a - 2 = 3a - 4, x \in I\}$

3. Simplify:

    (a) $3(4r - 4r) + 2(5r + 2s)$

    (b) $8(2r + 3t) - 10(-r - 2t)$

    (c) $5(6 + x + 2y) + 2[3 + 4x + 3y - (3x + 2y)]$

    (d) $\dfrac{7a}{2} - \dfrac{a}{3}$

    (e) $\dfrac{5y}{4x} + \dfrac{7y}{2x}$

4. Which of the following numerals represent the reciprocal of $-\frac{2}{3}$?

    (a) $\frac{2}{3}$ (e) $-1(-\frac{2}{3})$

    (b) $\dfrac{-\frac{2}{3}}{-1}$ (f) $\frac{3}{2}$

    (c) $\dfrac{-6}{4}$ (g) $\dfrac{6}{-4}$

    (d) $1 \div (-\frac{2}{3})$ (h) $\dfrac{3}{-2}$

5. If $x > 0$ and $y < 0$ state whether each of the following is positive or negative.

    (a) $x y$ (d) $-x$

(b) x^2 (e) $|y|$

(c) $-(y)^2$ (f) $x + y^2$

6. Factor each of the following:

   (a) $8p^2 + 26p + 21$ if one factor is $(4p + 7)$

   (b) $28 + 3x - 18x^2$ if one factor is $(4 - 3x)$

   (c) $5a^2 - 20$

   (d) $x^2 - 10^2$

   (e) $a^2 - 5ab - 6b^2$

7. Perform the indicated operation in each case:

   (a) Subtract 53 from $x + 21$.

   (b) Add $x + 3$ to the sum of x and $2x - 4$.

   (c) Multiply the sum of 4 and 5 by the difference
       between $2x$ and 3.

   (d) Divide $x^2 - 6x + 8$ by $x - 4$ by factoring divisor and
       dividend fully and dividing out like factors.

   (e) Divide x by 3 and add the quotient to the product of
       $2x$ and 5.

8. (a) Find the sum of $\sqrt{2}$, $-3\sqrt{2}$, $4\sqrt{2}$

   (b) Find the value of $\sqrt{\dfrac{4x^2}{y^2}}$ when $x = 4$ and $y = -3$

   (c) Simplify $\sqrt{(-p)^2}$

   (d) Between what two consecutive integers does $\sqrt{7}$ lie ?

   (e) Express the repeating decimal $0.4\dot{3}\dot{1}$ as a rational number.

9. Graph the truth sets of each of the following inequalities:

   (a) $T = \{ x \mid -7 < -x \le 8, x \epsilon Q \}$

   (b) $S = r \mid -3 \le r < 5, r \epsilon R \}$

   (c) $\{ x \mid 4x - 3 \ge -12, x \epsilon R \}$

   (d) $\{ x \mid 3x - 2 > 4x + 3, x \epsilon R \}$

10. A room is 8 feet longer than it is wide. If both width and length
    are increased by 2 feet the area would be increased by 44 sq. ft.
    Find the dimensions of the room by using an algebraic equation.

SET OF SOLUTIONS TO EASTER EXAMINATION – SAMPLE 1

1. $N_o - \{0, 1, 2, 3 \ldots \}$

   $Q - \{\frac{a}{b} \mid a, b \in I, b \neq 0\}$

   $I - \{0, \pm 1, \pm 2, \pm 3 \ldots\}$

2. (a) $\{x \mid x - 3 < 2, x \in N\}$

   $x - 3 < 2$

   $x < 5$

   The graph represents the solution set.

   (b) $\{x \mid x + 3 - 8, x \in N\}$

   $x + 3 - 8$

   $x - 5$

   The graph does not represent the solution set because it shows a signed 5 as the root whereas the members of N are unsigned.

   (c) $\{a \mid 5a - 2 - 3a - 4, x \in I\}$

   $5a - 2 - 3a - 4$

   $2a - -2$

   $a - -1$

   The graph does not represent the solution set.

3. (a) $3(4r - 4r) + 2(5r + 2s)$

   $- 10r + 4s$

   Note that the first term is zero because 4r and −4r are opposites.

   (b) $8(2r + 3t) - 10(-r - 2t)$

   $- 16r + 24t + 10r + 20t$

   $- 26r + 44t$

   (c) $5(6 + x + 2y) + 2[3 + 4x + 3y - (3x + 2y)]$

   $- 30 + 5x + 10y + 2[3 + 4x + 3y - 3x - 2y]$

   $- 30 + 5x + 10y + 6 + 8x + 6y - 6x - 4y$

   $- 36 + 7x + 12y$

   (d) $\frac{7a}{2} - \frac{a}{3}$

   $\frac{21a - 2a}{6}$

   $\frac{19a}{6}$

(e) $\dfrac{5y}{4x} + \dfrac{7y}{2x}$

$= \dfrac{5y + 14y}{4x}$

$= \dfrac{19y}{4x}$

4. (a) $\left(\dfrac{2}{3}\right)\left(-\dfrac{2}{3}\right)$

$= -\dfrac{4}{9}$, and $-\dfrac{4}{9} \neq 1$

(b) $\dfrac{\left(-\dfrac{2}{3}\right)\left(-\dfrac{2}{3}\right)}{-1}$

$= -\dfrac{4}{9}$, and $-\dfrac{4}{9} \neq 1$

(c) $\left(\dfrac{-6}{4}\right)\left(\dfrac{-2}{3}\right)$

$= 1$ ∴ (c) represents the reciprocal

(d) $\dfrac{1}{\left(-\dfrac{2}{3}\right)}\left(-\dfrac{2}{3}\right)$

$= 1$ ∴ (d) represents the reciprocal

(e) $(-1)\left(-\dfrac{2}{3}\right)\left(-\dfrac{2}{3}\right)$

$= \dfrac{-4}{9}$, and $\dfrac{-4}{9} \neq 1$

(f) $\left(\dfrac{3}{2}\right)\left(\dfrac{-2}{3}\right)$

$= -1$, and $-1 \neq 1$

(g) $\left(\dfrac{6}{-4}\right)\left(-\dfrac{2}{3}\right)$

$= 1$ ∴ (g) represents the reciprocal

(h) $\left(\dfrac{3}{-2}\right)\left(-\dfrac{2}{3}\right)$

$= 1$ ∴ (h) represents the reciprocal

5.   (Note that x > 0 and y < 0)

(a) xy < 0 because x and y differ in sign.

(b) $x^2 > 0$ because any number (of R) is positive whensquared.

(c) $-(y)^2 < 0$ because the negative of y^2 is taken. Because $y^2 > 0$ the final result is negative.

(d) -x < 0 because x is positive.

(e) | y | > 0 by definition of absolute value.

(f) $x + y^2 > 0$ because it is the sum of two positive numbers.

6.   (a) $8p^2 + 26p + 21$
      $= (4p + 7)(2p + 3)$

(b) $28 + 3x - 18x^2$
      $= (4 - 3x)(7 + 6x)$

(c) $5a^2 - 20$
      $= 5(a^2 - 4)$
      $= 5(a + 2)(a - 2)$

(d) $x^2 - 10^2$
      $= (x + 10)(x - 10)$

(e) $a^2 - 5ab - 6b^2$
      $= (a - 6b)(a + b)$

7.   (a) $(x + 21) - 53$
      $= x + 21 - 53$
      $= x - 32$

(b) $2x - 4 + x + (x + 3)$
      $= 2x - 4 + x + x + 3$
      $= 4x - 1$

(c) $(4 + 5)(2x - 3)$
      $= 9(2x - 3)$
      $= 18x - 27$

(d) $\dfrac{x^2 - 6x + 8}{x - 4}$

$= \dfrac{(x \overset{1}{-} 4)(x - 2)}{x \underset{1}{-} 4}$

$= x - 2$

(e) $\dfrac{x}{3} + (2x)(5)$

$= \dfrac{x}{3} + 10x$

$= \dfrac{x + 30x}{3}$

$= \dfrac{31x}{3}$

8.   (a) $\sqrt{2} + (-3\sqrt{2}) + 4\sqrt{2}$

$= \sqrt{2} - 3\sqrt{2} + 4\sqrt{2}$

$= 2\sqrt{2}$

(b) $\sqrt{\dfrac{4x^2}{y^2}}$

$= \sqrt{\dfrac{4(4)(4)}{(-3)(-3)}}$

$= \sqrt{\dfrac{64}{9}}$

$= \dfrac{8}{3} = 2\frac{2}{3}$

(c) $\sqrt{(-p)^2}$

$= \sqrt{p^2}$

$= |p|$

(d) $\sqrt{7}$ lies between 2 and 3.

(e) let $x = 0.4\dot{3}\dot{1}$ (1)

then $1000x = 431.4\dot{3}\dot{1}$ (2)

Subtract (1) from (2)

$999x = 431$

$x = \dfrac{431}{999}$

$\therefore 0.4\dot{3}\dot{1}$ expressed in rational form is $\dfrac{431}{999}$.

9. (a) $T = \{x \mid -7 < -x \leq 8, x \in Q\}$

$-7 < -x \leq 8$ Note in the graph that there are

$\therefore 7 > x \geq -8$ unrecorded "holes" in the diagram.

These are filled by the irrationals.

(b) $S = \{r \mid -3 \leq r < 5, r \in R\}$

(c) $\{x \mid 4x - 3 \geq -12, x \in R\}$

$4x - 3 \geq -12$

$\Longleftrightarrow 4x \geq -9$

$\Longleftrightarrow x \geq \dfrac{9}{4}$

(d) $\{x \mid 3x - 2 > 4x + 3, x \in R\}$

$3x - 2 > 4x + 3$

$\Longleftrightarrow -x > 5$

$\Longleftrightarrow x < -5$

10.   Let the width in feet be x

Then the length is x + 8

The new width is x + 2

The new length is x + 8 + 2

$(x + 2)(x + 10) - x(x + 8) = 44$

$\Longleftrightarrow x^2 + 12x + 20 - x^2 - 8x = 44$

$\Longleftrightarrow 4x = 24$

$\Longleftrightarrow x = 6$

The dimensions of the room are width 6'

length 14'

Check:

$$8 \times 16 = 128$$
$$6 \times 14 = \underline{\ 84}$$
$$\text{Diff} \ \underline{\ 44} \ \checkmark$$

EASTER EXAMINATION – SAMPLE 2
(Suggested time 2 hr.)

1. Match the proper answers in column 2 to the questions in column 1. (Note that there is NOT a one-to-one correspondence between the elements in the two columns.)

Column I	Column II
(a) $5 - 2 \times 3 + 7$	A – 37
(b) The value of x^2 where $x = 4$	B – 16
(c) $8 - 8 \div 2 + 0 \times 3$	C – 0
(d) The value of abc where $a = \frac{1}{2}$, $b = 8$	D – 4
$c = 0$	E – –3
(e) $(-5b + 17) + 5(b + 2a) - 10(a + 2)$	F – 6

2. In each of the following, insert brackets to make the statement true.

   (a) $5 \times 3 - 2 = 5$

   (b) $8 - 2 + 3 \times 4 = 36$

   (c) $8 \div 4 + 4 + 1 = 2$

   (d) $0 - 5 + 3 \times 8 = -64$

3. Simplify:

   (a) $4(3x + 5) + x(2x - 2) - 4x^2$

   (b) $-5p(12x - 2y) - 2p(x + 3y)$

   (c) $\frac{1}{4}(-8a)(-8b)c - \frac{1}{3}(3b)(-2c)(-3a) - \frac{1}{5}(-15ab)(-c)$

   (d) $\frac{x^2 + z}{x^2} - \frac{x^3 + z}{x^2}$

   (e) $\frac{3}{x + y} + \frac{2}{x + y}$

4. Name the numerals in the following list which represent the opposite of 4:

(a)   –4

(b)   4

(c) $\dfrac{1}{-4}$

(d)   (2)(–1)(–2)

(e)   (2)(1)(–2)

(f)   (–1)(4)(–1)

(g) $\dfrac{-4}{-1}$

(h)   (–2)(–2)

5. Write each of the following pairs of numbers in order starting with the smaller:

(a) –88, –89

(b) $\frac{8}{5}$, $\frac{7}{6}$

(c) 2(3) –8, 2(3 – 8)

(d) | 3 | , | –4 |

(e) $a^2 + 1$, 0

6. Factor numerator and denominator of each fraction and divide out like factors:

(a) $\dfrac{a^2 - 7a + 12}{a^2 - 5a + 6}$ x $\dfrac{a^2 - 4a + 3}{a^2 - 5a + 6}$ x $\dfrac{a^2 - 3a + 2}{a^2 - 6a + 5}$

(b) $\dfrac{2mn + 18m}{10mn + 90m}$ x $\dfrac{3m + 9}{6mn + 18n}$

7. Solve each of the following equations and check your answer:

(a) $5x + 3 = 15 - x$, $x \in N$

(b) $2x^2 = (x + 1)^2 + (x + 3)^2$, $x \in Q$

(c) $3(a - 1) - \dfrac{a - 4}{3} = 0$, $x \in Q$

(d) $12a - 2(4a + 1) = a$, $x \in Q$

8. The following is a well-tried method of "closing in" on the square root of a number:

1. Given a number "a" whose square root is to be found;

2. Choose *any* number (less than "a" because the root of a number is less than the number, $a \geq 1$) .

3. Divide this number into "a".

4. Average the quotient found in 3 and the number chosen in 2.

5. Divide *this* number into "a".

6. Average the quotient found in 5 with the quotient found in 3.

7. Repeat as often as you like.

Read the above directions carefully and thereby find a square root of 93 correct to one decimal place.

9. Equivalent equations or inequalities are those that have the same solution sets. Using this definition determine which of the following pairs are equivalent and which are not equivalent. $x \in Q$

(a) $5x - 1 < -11$

$-4(x + 1) + 7 > -3x + 5$

(b) $\frac{x}{5} - 2 > 1$

$2 + \frac{x}{3} > 7$

(c) $3(x - 2) \leq 2x + 8$

$2 - 3x \leq -7$

(d) $3x + 4 < -8$

$\frac{x - 1}{5} > x + 3$

10. One square has a side 2 inches longer than another square. If the areas differ by 24 sq. in. find the length of the side of the smaller square.

SET OF SOLUTIONS TO EASTER EXAMINATION – SAMPLE 2

1. (a) $5 - 2 \times 3 + 7$

$= 5 - 6 + 7$

$= 6$ match with F

(b) when $x = 4$

$x^2 = 16$ match with B

(c) $8 - 8 \div 2 + 0 \times 3$

$= 8 - 4$

$= 4$ match with D

(d) when $a = \frac{1}{2}$, $b = 8$, $c = 0$

$abc = 0$ match with C

(e) $(-5b + 17) + 5(b + 2a) - 10(a + 2)$

$= -5b + 17 + 5b + 10a - 10a - 20$

$= -3$ match with E

2. (a) $5 \times (3 - 2) = 5$

  (b) $(8 - 2 + 3) \times 4 = 36$

  (c) $8 \div (4 + 4) + 1 = 2$

  (d) $0 - (5 + 3) \times 8 = -64$

3. (a) $4(3x + 5) + x(2x - 2) - 4x^2$

    $= 12x + 20 + 2x^2 - 2x - 4x^2$

    $= -2x^2 + 10x + 20$

  (b) $-5p(12x - 2y) - 2p(x + 3y)$

    $= -60px + 10py - 2px - 6py$

    $= -62px + 4py$

  (c) $\frac{1}{4}(-8a)(-8b)c - \frac{1}{3}(3b)(-2c)(-3a) - \frac{1}{5}(-15ab)(-c)$

    $= 16abc - 6abc - 3abc$

    $= 7abc$

  (d) $\dfrac{x^2 + z}{x^2} - \dfrac{x^3 + z}{x^2}$

    $= \dfrac{x^2 + z - x^3 - z}{x^2}$

    $= \dfrac{x^2 - x^3}{x^2}$

    $= \dfrac{\overset{1}{\cancel{x^2}}(1 - x)}{\underset{1}{\cancel{x^2}}}$

    $= 1 - x$

4. The opposite of 4 is −4

    (a) −4   represents the opposite of 4.

    (b) 4   does not represent the opposite of 4.

    (c) $\dfrac{1}{-4} = -\dfrac{1}{4}$ does not represent the opposite of 4.

    (d) $(2)(-1)(-2)$
       $= 4$ does not represent the opposite of 4.

    (e) $(2)(1)(-2)$
       $= -4$ represents the opposite of 4.

    (f) $(-1)(4)(-1)$
       $= 4$ does not represent the opposite of 4.

    (g) $\dfrac{-4}{-1}$
       $= 4$ does not represent the opposite of 4.

    (h) $(-2)(-2)$
       $= 4$ does not represent the opposite of 4.

5.   (a) $-89 < -88$

    (b) $\dfrac{7}{8} < \dfrac{8}{9}$ because in comparing by cross multiplication
                                 $63 < 64.$

    (c) $2(3) - 8 = -2$
       $2(3 - 8) = -10$
    $\therefore\ 2(3 - 8) < 2(3) - 8$

    (d) $|\,3\,| < |\,-4\,|$ by definition

    (e) $0 < (a^2 + 1)$

6.   (a) $\dfrac{a^2 - 7a + 12}{a^2 - 5a + 6} \times \dfrac{a^2 - 4a + 3}{a^2 - 5a + 6} \times \dfrac{a^2 - 3a + 2}{a^2 - 6a + 5}$

$$= \frac{(a-4)\overset{1}{\cancel{(a-3)}}}{\underset{1}{\cancel{(a-3)}}\,\underset{1}{\cancel{(a-2)}}} \times \frac{\overset{1}{\cancel{(a-3)}}\,\overset{1}{\cancel{(a-1)}}}{\underset{1}{\cancel{(a-3)}}(a-2)} \times \frac{\overset{1}{\cancel{(a-2)}}(a-1)}{(a-5)\,\underset{1}{\cancel{(a-1)}}}$$

$= \dfrac{(a-4)(a-1)}{(a-2)(a-5)}$ either this, or multiplied out, is a satisfactory form.

(b) $\dfrac{2mn + 18m}{10mn + 90m} \times \dfrac{3m + 9}{6mn + 18n}$

$= \dfrac{\overset{1}{\cancel{2m}} \overset{1}{\cancel{(n + 9)}}}{\underset{5}{\cancel{10m}} \underset{1}{\cancel{(n + 9)}}} \times \dfrac{\overset{1}{\cancel{3}} \overset{1}{\cancel{(m + 3)}}}{\underset{2}{\cancel{6n}} \underset{1}{\cancel{(m + 3)}}}$

$= \dfrac{1}{10n}$

7. (a) $5x + 3 = 15 - x$

$\Longleftrightarrow \quad 6x = 12$

$\Longleftrightarrow \quad x = 2$

Check:

L.H.N. $= 10 + 3 = 13$

R.H.N. $= 15 - 2 = 13$

∴ 2 is the root of the equation.

(b) $2x^2 = (x + 1)^2 + (x + 3)^2$

$\Longleftrightarrow 2x^2 = x^2 + 2x + 1 + x^2 + 6x + 9$

$\Longleftrightarrow -8x = 10$

$\Longleftrightarrow \quad x = -\dfrac{5}{4}$

Check:

L.H.N. $= 2\left(\dfrac{-5}{4}\right)^2$

$\qquad = \dfrac{25}{8}$

R.H.N. $= \left(\dfrac{-5}{4} + 1\right)^2 + \left(\dfrac{-5}{4} + 3\right)^2$

$\qquad = \left(-\dfrac{1}{4}\right)^2 + \left(\dfrac{7}{4}\right)^2$

$\qquad = \dfrac{1}{16} + \dfrac{49}{16}$

$\qquad = \dfrac{25}{8}$

∴ $-\dfrac{5}{4}$ is the root of the equation.

(c) $3(a-1) - \dfrac{a-4}{3} = 0$

Multiply each term by 3.

$\Longleftrightarrow 9a - 9 - a + 4 = 0$

$\Longleftrightarrow \qquad\qquad 8a = 5$

$\Longleftrightarrow \qquad\qquad a = \dfrac{5}{8}$

Check:

$L.H.N. = 3\left(\dfrac{5}{8} - 1\right) - \dfrac{\dfrac{5}{8} - 4}{3}$

$\qquad = 3\left(-\dfrac{3}{8}\right) - \dfrac{5 - 32}{24}$

$\qquad = \dfrac{-9}{8} + \dfrac{9}{8}$

$\qquad = 0 = R.H.N.$

$\dfrac{5}{8}$ is the root of the equation.

(d) $12a - 2(4a + 1) = a$

$\Longleftrightarrow 12a - 8a - 2 \quad = a$

$\Longleftrightarrow \qquad\qquad 3a = 2$

$\Longleftrightarrow \qquad\qquad a = \frac{2}{3}$

Check:

$L.H.N. = 12\left(\dfrac{2}{3}\right) - 2\left(\dfrac{8}{3} + 1\right)$

$\qquad = 8 - \dfrac{22}{3}$

$\qquad = \dfrac{2}{3} = R.H.N.$

$\therefore \dfrac{2}{3}$ is the root of the equation.

8. Following the directions let us choose 6.

$$93 \div 6 = 15.50$$

Average 6 and 15.50 is $\dfrac{6 + 15.50}{2} = 10.75$

$$93 \div 10.75 = 8.65$$

Average 8.65 and 10.75 is $\dfrac{8.65 + 10.75}{2} = 9.70$

$$93 \div 9.70 = 9.59$$

Average 9.59 and 9.70 is $\dfrac{9.59 + 9.70}{2} = 9.65$

$$93 \div 9.65 = 9.64$$

The required root to the proper degree of accuracy is 9.6

9. (a) $5x - 1 < -11$ $-4(x + 1) + 7 > -3x + 5$

   $\Longleftrightarrow 5x < -10$ $\Longleftrightarrow -4x - 4 + 7 > -3x + 5$

   $\Longleftrightarrow x < -2$ $\Longleftrightarrow \quad\quad -x > -2$

   $\phantom{\Longleftrightarrow x < -2}$ $\Longleftrightarrow \quad\quad\quad x < 2$

These two inequalities are NOT equivalent.

(b) $\dfrac{x}{5} - 2 > 1$ $2 + \dfrac{x}{3} > 7$

   $\Longleftrightarrow x - 10 > 5$ $\Longleftrightarrow 6 + x > 21$

   $\Longleftrightarrow \quad x > 15$ $\Longleftrightarrow \quad\quad x > 15$

These inequalities are equivalent.

(c) $3(x - 2) \leq 2x + 8$ $2 - 3x \leq -7$

   $\Longleftrightarrow 3x - 6 \leq 2x + 8$ $\Longleftrightarrow -3x \leq -9$

   $\Longleftrightarrow \quad\quad x \leq 14$ $\Longleftrightarrow \quad\quad x \geq 3$

These inequalities are NOT equivalent.

(d) $3x + 4 < -8$ $\dfrac{x - 1}{5} > x + 3$

   $\Longleftrightarrow 3x < -12$ $\Longleftrightarrow x - 1 > 5x + 15$

   $\Longleftrightarrow \quad x < -4$ $\Longleftrightarrow -4x > 16$

   $\phantom{\Longleftrightarrow x < -4}$ $\Longleftrightarrow \quad\quad x < -4$

These inequalities are equivalent.

10. Let the side of the smaller square in inches be x

Then the side of the larger square is x + 2

$$(x + 2)^2 - x^2 = 24$$
$$\Longleftrightarrow x^2 + 4x + 4 - x^2 = 24$$
$$\Longleftrightarrow \qquad\qquad 4x = 20$$
$$\Longleftrightarrow \qquad\qquad x = 5$$

The side of the smaller square is 5

Check:

$$7^2 = 49$$
$$5^2 = 25$$
$$\text{Diff} = 24 \quad \sqrt{}$$

EASTER EXAMINATION – SAMPLE 3
(Suggested time 2hr.)

1. If x belongs to the set $\{1, 2, 3, 4, 5, 6, 7, 8, 9\}$

(a) Solve by inspection x + 5 = 4

(b) Solve by inspection   2x = x + 7

(c) Solve by inspection 3x − 4 = 8

2. Simplify:

(a) −4 (r − 3) + 5 (r + 3)

(b) −7a − 3(−2a + 5b) − b

(c) 8(−2x) + 2 (−3x + 5) − 2x [4 − 3 (−2 −2x)]

(d) $\dfrac{7b}{8} - \dfrac{5b}{8}$

(e) $\dfrac{2b}{3(a + 1)} + \dfrac{5c}{3(a + 1)}$

3. From the following list choose the numerals that represent the opposite of $\dfrac{2 + 5}{-3 + 7}$:

(a) $(-1) \left(\dfrac{2 + 5}{-3 + 7}\right)$ \qquad\qquad (d) $\dfrac{-3 + 7}{-2 - 5}$

(b) $\dfrac{-2 - 5}{3 - 7}$　　　　　　　**(e)** $\left(\dfrac{2 + 5}{-3 + 7}\right) \div (-1)$

(c) $\dfrac{2 + 5}{3 + 7}$　　　　　　　**(f)** $\dfrac{2 + 5}{(-1)(-3 + 7)}$

4. Evaluate each of the following if a = 3, b = 0, c = –4:

   (a) $b^2 - 4ac$

   (b) $2b(3a - 4c)$

   (c) $3a^2 c^3$

   (d) $\dfrac{5bc}{a}$

   (e) $(a + b + c) - (a + b + c) + (a + b + c)$

5. Factor:

   (a) $3abc + 27ab^2$

   (b) $a^2 + 14a + 49$

   (c) $169b^2 - 121x^2$

   (d) $r^3 - 3r^2 - 28r$

   (e) $(p + q)x + (p + q)y$

6. Solve each of the following equations and check your answer:

   (a) $(x + 4)(x - 2) = x(x - 2)$

   (b) $(x - 2)(x + 4) - (x + 2)(x + 4) = -24$

   (c) $\frac{1}{2}x + \frac{2}{3} = \frac{1}{2}x - \frac{1}{3}$

   (d) $\dfrac{x + 5}{2} = \dfrac{x - 5}{3}$

7. (a) Between which two consecutive integers does $\sqrt{183}$ lie?

   (b) Estimate $\sqrt{105}$ correct to one place of decimals.

   (c) By factoring find $\sqrt{x^2 - 2xy + y^2}$.

   (d) Express the repeating decimal $0.3\dot{8}$ as a rational number.

8. Graph the following inequalities:

   (a) $3a - 2 > a + 8$, $a \in I$.

(b) $2(x + 4) - 1 \geq x + 17$, $x \in R$.

(c) $-r + 3 \leq -4r - 6$, $x \in R$.

(d) $x(x + 5) - 1 > x^2 + 8(x + 1)$, $x \in R$

9. A boy carries 20 coins in his pocket in half dollars and dimes. He counts them, and finds that he has $3.60. How many of each kind of coin has he?

10. A girl is twice as old as her brother. Four years ago she was four times as old as her brother. Find their present ages.

SET OF SOLUTIONS TO EASTER EXAMINATION – SAMPLE 3

1. (a) $x + 5 = 4$

   ∴ the solution set is $\{1\}$

   (b) $2x = x + 7$

   ∴ the solution set is $\{7\}$

   (c) $3x - 4 = 8$

   ∴ the solution set is $\{4\}$

2. (a) $-4(r - 3) + 5(r + 3)$

   $= -4r + 12 + 5r + 15$

   $= r + 27$

   (b) $-7a - 3(-2a + 5b) - b$

   $= -7a + 6a - 15b - b$

   $= -a - 16b$

   (c) $8(-2x) + 2(-3x + 5) - 2x[4 - 3(-2 - 2x)]$

   $= -16x - 6x + 10 - 2x[4 + 6 + 6x]$

   $= -22x + 10 - 8x - 12x - 12x^2$

   $= -12x^2 - 42x + 10$

   (d) $\dfrac{7b}{8} - \dfrac{5b}{8}$

   $= \dfrac{2b}{8}$

   $= \dfrac{b}{4}$

(e) $\dfrac{2b}{3(a+1)} + \dfrac{5c}{3(a+1)}$

$= \dfrac{2b+5c}{3(a+1)}$

3. The opposite to $\dfrac{2+5}{-3+7}$ is $-\dfrac{2+5}{-3+7} = -\dfrac{7}{4}$

(a) $(-1)(\dfrac{2+5}{-3+7})$

$= -\dfrac{7}{4}$ This represents the opposite to the given number.

(b) $\dfrac{-2-5}{3-7}$

$= \dfrac{-7}{-4}$

$= \dfrac{7}{4}$ This does NOT represent the opposite to the given number.

(c) $\dfrac{2+5}{3+7}$

$= \dfrac{7}{10}$ This does NOT represent the opposite to the given number.

(d) $\dfrac{-3+7}{-2-5}$

$= \dfrac{4}{-7}$ This does NOT represent the opposite to the given number.

(e) $(\dfrac{2+5}{-3+7}) \div (-1)$

$= -\dfrac{7}{4}$ This represents the opposite to the given number.

(f) $\dfrac{2+5}{(-1)(-3+7)}$

$= -\dfrac{7}{4}$ This represents the opposite to the given number.

4. (a) $b^2 - 4ac$

$= 0 - 4(3)(-4)$

$= 48$

(b) $2b (3a - 4c)$

    $= 0$

(c) $3a^2c^3$

    $= 3(3)(3)(-4)(-4)(-4)$

    $= -1728$

(d) $\dfrac{5bc}{a}$

    $= 0$

(e) $(a + b + c) - (a + b + c) + (a + b + c)$

    $= a + b + c$

    $= 3 + 0 - 4$

    $= -1$

5. (a) $3abc + 27ab^2$

    $= 3ab (c + 9b)$

  (b) $a^2 + 14a + 49$

    $= (a + 7)^2$

  (c) $169b^2 - 121x^2$

    $= (13b + 11x)(13b - 11x)$

  (d) $r^3 - 3r^2 - 28r$

    $= r (r^2 - 3r - 28)$

    $= r (r - 7)(r + 4)$

  (e) $(p + q) x + (p + q) y$

    $= (p + q)(x + y)$

6. (a) $(x + 4)(x - 2) = x(x - 2)$

    $\Longleftrightarrow x^2 + 2x - 8 = x^2 - 2x$

    $\Longleftrightarrow \qquad 4x = 8$

    $\Longleftrightarrow \qquad x = 2$

Check:

L.H.N. $= (2 + 4)(2 - 2)$ R.H.N. $= 2(2 - 2)$

    $= 0$ $= 0$

∴ 2 is the root of the equation.

(b) $(x - 2)(x + 4) - (x + 2)(x + 4) = -24$

$\Longleftrightarrow x^2 + 2x - 8 - x^2 - 6x - 8 \quad = -24$

$\Longleftrightarrow \qquad\qquad\qquad -4x = -8$

$\Longleftrightarrow \qquad\qquad\qquad x = 2$

Check:

L.H.N. $= (2 - 2)(2 + 4) - (2 + 2)(2 + 4)$

$\qquad = -(4)(6)$

$\qquad = -24 = $ R.H.N.

$\therefore$ 2 is the root of the equation.

(c) $\frac{1}{2}x + \frac{2}{3} = \frac{3}{2}x - \frac{1}{3}$

Multiply each term by 6.

$\Longleftrightarrow 3x + 4 = 9x - 2$

$\Longleftrightarrow \quad = 6x = -6$

$\Longleftrightarrow \quad x = 1$

Check:

L.H.N. $= \frac{1}{2}(1) + \frac{2}{3}$ R.H.N. $= \frac{3}{2}(1) - \frac{1}{3}$

$\qquad = 1\frac{1}{6}$ $\qquad\qquad = 1\frac{1}{6}$

$\therefore$ 1 is the root of the equation.

(d) $\dfrac{x + 5}{2} = \dfrac{x - 5}{3}$

Multiply each side by 6

$\Longleftrightarrow 3x + 15 = 2x - 10$

$\Longleftrightarrow \quad x = -25$

Check:

L.H.N. $= \dfrac{-25 + 5}{2} = -10$

R.H.N. $= \dfrac{-25 - 5}{3} = -10$

$\therefore$ -25 is the root of the equation.

7. (a) $\sqrt{183}$ lies between 13 and 14

(b) First estimate $10.1^2 = 102.01$
Second estimate $10.2^2 = 104.04$
Third estimate $10.3^2 = 106.09$

Estimate $\sqrt{105} = 10.2$ correct to one decimal place.

(c) $\sqrt{x^2 - 2xy + y^2}$

$= \sqrt{(x - y)^2}$

$= |x - y|$

(d)   let $x = 0.\dot{3}\dot{8}$ (1)

then $100x = 38.\dot{3}\dot{8}$ (2)

Subtract (1) from (2)

$99x = 38$

$\therefore x = \dfrac{38}{99}$

and $0.\dot{3}\dot{8} = \dfrac{38}{99}$ as a rational number

8. (a) $3a - 2 > a + 8$

$\iff 2a > 10$

$\iff a > 5$

$\overset{\text{-1 0 '1 '2 '3 '4 '5 '6 '7 '8}}{\underset{\text{I-line}}{\longleftrightarrow}}$

(b) $2(x + 4) - 1 \geq x + 17$

$\iff 2x + 8 - 1 \geq x + 17$

$\iff x \geq 10$

$\overset{\text{'7 '8 '9 '10 '11 '12 '13 '14 '15 '16}}{\underset{\text{R-line}}{\longleftrightarrow}}$

98

(c) $-r + 3 \leq -4r - 6$

  $\Longleftrightarrow 3r \leq -9$

  $\Longleftrightarrow r \leq -3$

$$\underset{\text{-8 -7 -6 -5 -4 -3 -2 -1 0 +1}}{\xrightarrow{\hspace{5cm}}} \text{R-line}$$

(d) $x(x + 5) - 1 > x^2 + 8(x + 1)$

  $\Longleftrightarrow x^2 + 5x - 1 > x^2 + 8x + 8$

  $\Longleftrightarrow \qquad -3x > 9$

  $\Longleftrightarrow \qquad x < -3$

$$\underset{\text{-8 -7 -6 -5 -4 -3 -2 -1 0 +1}}{\xrightarrow{\hspace{5cm}}} \text{R-line}$$

9.   Let the number of half dollars be x

  Then the number of dimes is $20 - x$

  $50x + 10(20 - x) = 360$

$\Longleftrightarrow 50x + 200 - 10x = 360$

$\Longleftrightarrow \qquad 40x = 160$

$\Longleftrightarrow \qquad x = 4$

  The number of half dollars is 4

  The number of dimes is 16

  Check

$$50 \times 4 = 200$$
$$10 \times 16 = \underline{160}$$
$$\text{Total} \quad \underline{360} \ \surd$$

10.   Let the girl's present age be x

  Her brother's present age is $\frac{x}{2}$

  Ages of girl and brother four years ago was $x-4$ and $\frac{x}{2} - 4$ resp.

$$x - 4 = 4(\frac{x}{2} - 4)$$

$$x - 4 = 2x - 16$$

$$-x = -12$$

$$x = 12$$

The girl's present age is 12

Her brother's present age is 6

Check

> Girl's age 4 years ago was 8
>
> Boy's age 4 years ago was 2
>
> and $8 = 4 \times 2$ √

JUNE EXAMINATION SAMPLE
(Suggested time – 2 hr.)

1.   (a) Add:

$$3xy - 2yz + zx$$
$$-4xy + yz - 2zx$$
$$\underline{-xy - yz - zx}$$

    (b) Subtract:

$$2a + 4b - 2c + d - e$$
$$\underline{-a - 5b + c + d + e}$$

2. Simplify:

   (a) $(3a + 4)(a - 1) + (a + 2)^2$

   (b) $(3a + 4)(a - 1) - (a + 2)^2$

   (c) $2(2x + y)(x - y) - 3(x + 2y)(3x - 4y)$

   (d) $2[3x - (2y + 5)] - (2x + y)^2$

3. Factor each expression. Simplify where necessary:

   (a) $4ax + 4bx$

   (b) $x^2 + 9x - 22$

   (c) $x^2 - 16$

   (d) $\dfrac{x^2 + 3x + 2}{x(x + 1)}$

(e) $\dfrac{x^2 - b^2}{x^2 - bx} \div \dfrac{x + b}{x}$

4. (a) If $a = -1$, $b = 2$, $c = -3$ evaluate $b^2 - 4ac$

   (b) If $x = 0$, $y = 3$, $z = 4$ evaluate $y^2(xy + yz)$

   (c) If $a = \frac{1}{2}$, $b = 2$, $c = -2$ evaluate $ba - (b + c) + abc$

   (d) If $x = a + b + c$, $y = -a -b -c$, $z = 2a - b^2 - c$

   evaluate $x + y + z$ in terms of a, b and c

5. Solve each of the following and check your answers:

   (a) $(x + 3)(x + 2) = x(x + 2) + 15$

   (b) $\dfrac{2a + 4}{3} = a - 1$

   (c) $\frac{1}{4}(x - 2) + \frac{1}{3}(x + 1) = 4$

6. Solve each of the following and plot on a suitable number line.

   (a) $(x + 3)(x + 2) = x(x + 2) + 15$, $x \in R$

   (b) $\{ x \mid 2x - 1 \geq 5 - x,\ x \in R \}$

   (c) $\{ x \mid x + 1 > 7 + 2x,\ x \in R \}$

7. Using the diagrams provided, perform the indicated
   constructions:

   (a) Bisect each angle of
   $\triangle ABC$
   Let bisectors of $\angle A$ and
   $\angle B$ meet in O
   From O draw OR $\perp$ AB
   with centre O and radius
   OR construct a circle.

   (b) Right bisect all sides
   of $\triangle XYZ$
   Let right bisector of XY
   and YZ meet in O
   With centre O and ra-
   dius OX draw a circle.

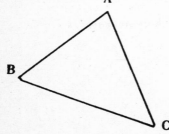

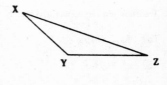

8.   (a) Construct an angle of 105°   (b) Construct an angle of 52½°

A   B   C

9.   (a)   (b)

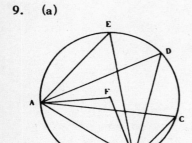   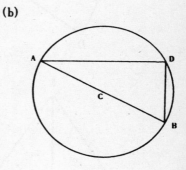

Measure ∠AFB, ∠E, D, C   If AB is a diameter, mea-
and draw possible conclu-   sure ∠ADB and draw
sions.   possible conclusion.

10.   In each of the following, name the congruent triangles, give
       the authority, and state three further equalities that are true:

(a)

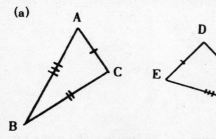

(b)

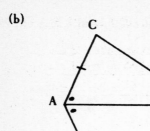

(c)

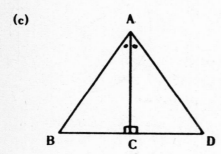

(d)

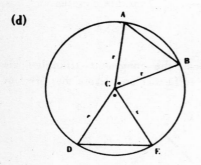

11. A grocer buys 50 cabbages and 40 bunches of carrots for $10.50. If the cabbages cost twice as much per head as the carrots per bunch, find the cost of each.

SET OF SOLUTIONS FOR JUNE EXAMINATION SAMPLE

1. (a)
$$3xy - 2yz + zx$$
$$-4xy + yz - 2zx$$
$$- xy - yz - zx$$
$$\overline{-2xy - 2yz - 2zx}$$

(b)
$$2a + 4b - 2c + d - e$$
$$-a - 5b + c + d + e$$
$$\overline{3a + 9b - 3c \qquad -2e}$$

2. (a) $(3a + 4)(a - 1) + (a + 2)^2$

   $= 3a^2 + a - 4 + a^2 + 4a + 4$

   $= 4a^2 + 5a$

   (b) $(3a + 4)(a - 1) - (a + 2)^2$

   $= 3a^2 + a - 4 - a^2 - 4a - 4$

   $= 2a^2 - 3a - 8$

   (c) $2(2x + y)(x - y) - 3(x + 2y)(3x - 4y)$

   $= 2(2x^2 - xy - y^2) - 3(3x^2 + 2xy - 8y^2)$

   $= 4x^2 - 2xy - 2y^2 - 9x^2 - 6xy + 24y^2$

   $= -5x^2 - 8xy + 22y^2$

   (d) $2[3x - (2y + 5)] - (2x + y)^2$

   $= 2[3x - 2y - 5] - (4x^2 + 4xy + y^2)$

   $= 6x - 4y - 10 - 4x^2 - 4xy - y^2$

3. (a) $4ax + 4bx$

   $= 4x(a + b)$

   (b) $x^2 + 9x - 22$

   $= (x + 11)(x - 2)$

   (c) $x^2 - 16$

   $= (x + 4)(x - 4)$

   (d) $\dfrac{x^2 + 3x + 2}{x(x + 1)}$

   $= \dfrac{(x + 2)\cancel{(x + 1)}}{x\cancel{(x + 1)}}$

   $= \dfrac{x + 2}{x}$

(e) $\dfrac{x^2 - b^2}{x^2 - bx} \div \dfrac{x + b}{x}$

$$= \dfrac{\overset{1}{\cancel{(x + b)}}\ \overset{1}{\cancel{(x - b)}}}{\underset{1}{\cancel{x}}\ \underset{1}{\cancel{(x - b)}}} \times \dfrac{\overset{1}{\cancel{x}}}{\underset{1}{\cancel{x + b}}}$$

$= 1$

4.   (a) $b^2 - 4ac$

$= (2)^2 - 4(-1)(-3)$

$= 4 - 12$

$= -8$

(b) $y^2 (xy + yz)$

$= (3)^2 (0 + 12)$

$= 108$

(c) $ba - (b + c) + abc$

$= 1 - 0 + (½)(2)(-2)$ Use ideas of reciprocals and

$= -1$ opposites here.

(d) $x + y + z$

$= 2a - b^2 - c$ by inspection and use of opposites in x and y

5.   (a) $(x + 3)(x + 2) = x(x + 2) + 15$

$\Longleftrightarrow x^2 + 5x + 6 = x^2 + 2x + 15$

$\Longleftrightarrow \qquad\qquad 3x = 9$

$\Longleftrightarrow \qquad\qquad x = 3$

   Check

   L.H.N. $= (3 + 3)(3 + 2)$ R.H.N. $= 3(3 + 2) + 15$

$\qquad = 6 \times 5 \qquad\qquad\qquad\qquad = 3 \times 5 + 15$

$\qquad = 30 \qquad\qquad\qquad\qquad\qquad = 30$

$\therefore$ 3 is the root of the equation

(b) $\dfrac{2a + 4}{3} = a - 1$

Multiply each term by 3

$\Longleftrightarrow 2a + 4 = 3a - 3$

$\Longleftrightarrow \quad -a = -7$

$\Longleftrightarrow \quad\quad a = 7$

Check

$$L.H.N. = \frac{14 + 4}{3} \qquad R.H.N. = 7 - 1$$

$$= 6 \qquad\qquad\qquad = 6$$

∴ 7 is the root of the equation

(c) $\frac{1}{4}(x - 2) + \frac{1}{3}(x + 1) = 4$

Multiply each term by 12

$\Longleftrightarrow 3(x - 2) + 4(x + 1) = 48$

$\Longleftrightarrow \quad 3x - 6 + 4x + 4 = 48$

$\Longleftrightarrow \quad\quad\quad\quad\quad 7x = 50$

$\Longleftrightarrow \quad\quad\quad\quad\quad x = \frac{50}{7} = 7\frac{1}{7}$

Check

$L.H.N. = \frac{1}{4}(7\frac{1}{7} - 2) + \frac{1}{3}(7\frac{1}{7} + 1)$

$\quad\quad = \frac{1}{4}(5\frac{1}{7}) + \frac{1}{3}(8\frac{1}{7})$

$\quad\quad = \frac{36}{28} + \frac{57}{21}$

$\quad\quad = \frac{9}{7} + \frac{19}{7}$

$\quad\quad = 4 = R.H.N.$

∴ $7\frac{1}{7}$ is the root of the equation

6. (a) $(x + 3)(x + 2) = x(x + 2) + 15, x \in R$

from 5(a) the correct root is 3

Fig.93.

(b) $\{x \mid 2x - 1 \geq 5 - x, x \in R\}$

$\quad\quad 2x - 1 \geq 5 - x$

$\Longleftrightarrow \quad 2x + x \geq 5 + 1$

$\Longleftrightarrow \quad\quad\quad x \geq 2$

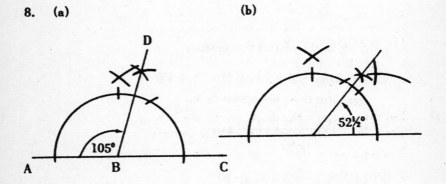

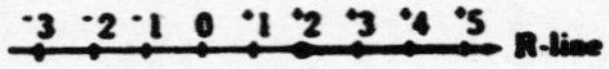
-3 -2 -1 0 +1 +2 +3 +4 +5
R-line

(c) $\{x \mid x + 1 > 7 + 2x,\ x \in R\}$

$$x + 1 > 7 + 2x$$

$$\Leftrightarrow \quad -x > 6$$

$$\Leftrightarrow \quad x < -6$$

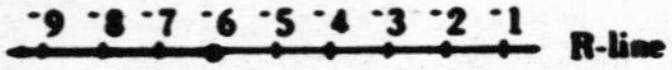
-9 -8 -7 -6 -5 -4 -3 -2 -1
R-line

7. (a)

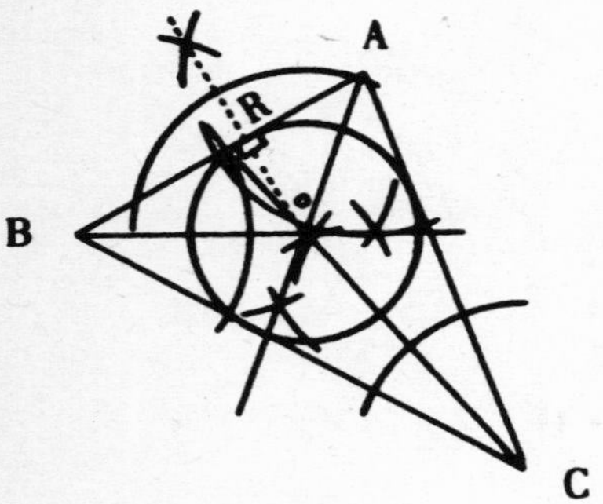
A
R
o
B
C

(b)

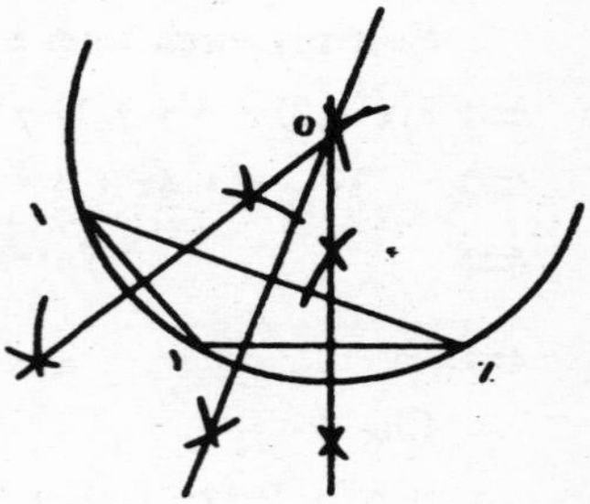
o

8. (a)

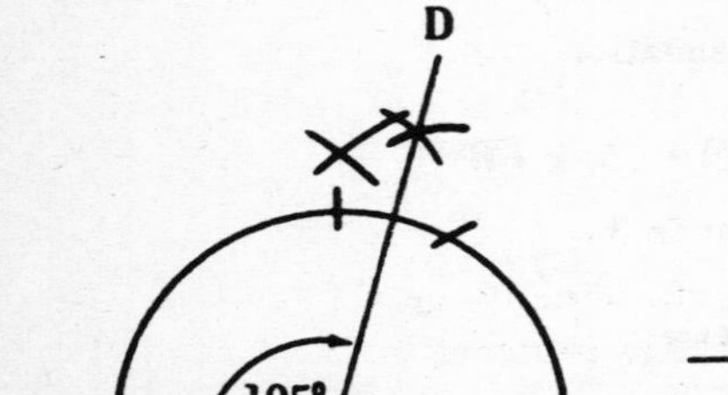
D
105°
A
B
C

(b)

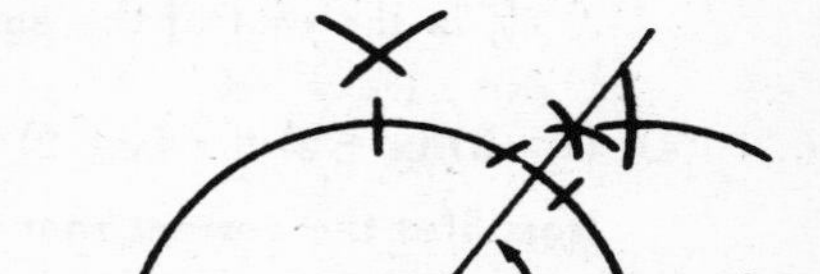
52½°

$105° = 90° + 15°$

$52\frac{1}{2}° = 45° + 7\frac{1}{2}°$

9. (a) $\angle AFB = 106°$

    $\angle AEB = 53°$

    $\angle ADB = 53°$

    $\angle ACB = 53°$

    It would appear that the angle at the circumference is half the corresponding sector angle.

    (b) $\angle ADB = 90°$

    It would appear that the angle in a semi-circle is a right angle.

10. (a) $\triangle ABC \equiv \triangle EDF$ (SSS)

    $\therefore \angle A = \angle E$

    $\angle B = \angle F$

    $\angle C = \angle D$

    (b) $\triangle ABC \equiv \triangle ABD$ (SAS)

    $\therefore \angle C = \angle D$

    $\angle CBA = \angle DBA$

    $CB = DB$

    (c) $\triangle ABC \equiv \triangle ADC$ (AAS)

    $\therefore \angle B = \angle D$

    $AB = AD$

    $BC = CD$

    (d) $\triangle ACB \equiv \triangle DCE$ (SAS)

    $\therefore \angle A = \angle D$

    $\angle E = \angle B$

    $AB = DE$

11. Let the cost of a bunch of carrots in cents be x

    Then the cost of a head of cabbage in cents is 2x

    $50 (2x) + 40x = 1050$

    $\Longleftrightarrow 100x + 40x = 1050$

    $\Longleftrightarrow 140x = 1050$

    $\Longleftrightarrow x = \dfrac{1050}{140}$

    $\Longleftrightarrow x = 7\frac{1}{2}$

The cost of a bunch of carrots is 7½ cents.

The cost of a head of cabbage is 15 cents.

Check

50 × .15   =   7.50

40 × .07½ =   3.00

Total   <u>$10.50</u> √